The Three Axial Ages

The Three Axial Ages

―

Moral, Material, Mental

JOHN TORPEY

Rutgers University Press

New Brunswick, Camden, and Newark, New Jersey, and London

Cataloging-in-Publication data is available from the Library of Congress.

A British Cataloging-in-Publication record for this book is available from the British Library.

Copyright © 2017 by John Torpey
All rights reserved

No part of this book may be reproduced or utilized in any form or by any means, electronic or mechanical, or by any information storage and retrieval system, without written permission from the publisher. Please contact Rutgers University Press, 106 Somerset Street, New Brunswick, NJ 08901. The only exception to this prohibition is "fair use" as defined by U.S. copyright law.

www.rutgersuniversitypress.org

The paper used in this publication meets the requirements of the American National Standard for Information Sciences—Permanence of Paper for Printed Library Materials, ANSI Z39.48-1992.

Manufactured in the United States of America

Contents

	Preface	vii
	Introduction	1
1.	The Moral Axial Age	7
2.	The Material Axial Age	33
3.	The Mental Axial Age	50
	Notes	79
	Index	97

Preface

This brief book arose from a certain disquiet about a recent discussion among (mainly) sociologists regarding the historical significance of the so-called Axial Age, the period in the middle centuries of the first millennium BCE when some of the world's major religions and intellectual developments emerged in China, India, the Middle East, and ancient Greece. The human significance of the birth of some of the world's most influential intellectual traditions was surely a matter that should command our attention and interest. This was especially true for me, given that one of the principal contributors to the discussion was a mentor of mine, the distinguished sociologist of religion Robert Bellah. Bob had always been an eclectic thinker, but also a profound one. His magnum opus on the Axial Age (*Religion in Human Evolution*) is astonishing in its range and depth of knowledge about the world's religious traditions. It also reads like an elaboration on Durkheim and Weber as guides to understanding the role of religion in human life. The book also had roots in a famous article of his from more than fifty years earlier, "Religious Evolution," whose stage-wise characterization of human religious development Bob attributed to his "early involvement with Marxism."[1] *Religion in Human Evolution* thus appeared in certain respects like a textbook based on sociology's "holy trinity" and reflecting their divergent ways of thinking about social life.

Yet there is no denying that the book, and the discussion of the Axial Age more generally, is profoundly idealist in its preoccupations. The defenders of the Axial Age thesis—that "man as we know him today" came into existence during that remarkable period two and a half millennia ago—take the position that the different intellectual traditions that arose at that time provide the basic modes of thinking of a large proportion of the world's peoples today. In view of the fact that we are talking about Buddhism, Hinduism, Confucianism, Taoism, Greek philosophy, and Judaism and its "Abrahamic" descendants, there is of course much to be said for this view; indeed, the "world religions" are so designated precisely because they shape the worldviews of substantial populations around the world. In part inspired by Michael Mann's writings, however, I believe we need to pay attention not just to the ideas but to various dimensions of particular epochs as well if we are to make sense of which factors predominate in defining particular historical conjunctures. Hence the three Axial Ages that I delineate in the pages that follow revolve principally, though of course by no means exclusively, around different historical "axes": the moral, the material, and the mental, respectively. What I mean by each of these terms will, I hope, become clear in the course of the exposition.

The present work also feeds off another sort of disquiet: that regarding the missing normative dimension of contemporary sociology. Having originally been an attempt to make sense of the modern world as it emerged from its predecessor agricultural societies, sociology has largely been reduced to the study of inequality, and is essentially "deconstructive" in character. That is, it is generally critical of current social arrangements of many kinds, but it shrinks from presenting more desirable social arrangements of its own. This non-normative stance contrasts notably with that of contemporary political science. The point can be made by comparing the place of "theory" in each field. Political theory, if hardly likely to vault a young scholar to the commanding heights of the discipline, is a perfectly legitimate—indeed, a venerable—part of the field, with a firm footing in the departments at leading universities. More importantly, political theory remains preoccupied with the nature and meaning of "the good life," as it has been ever since Plato and Aristotle. By contrast, in what is undoubtedly a negative legacy of Weber's strict Kantian distinction between the "is" and the "ought," "social theory" is about *understanding society* better, not about *making society better*. Hence the work of a normatively oriented scholar such as Jürgen Habermas, surely the most influential German social thinker of the second half of the twentieth century and often described as a sociologist, has little relevance to contemporary (American) sociology—especially by comparison with Pierre Bourdieu, an incisive thinker about matters of inequality but not a political or moral thinker of great import. The discussion about the Axial Age has thus been important because it has been a way to return moral considerations to the heart of sociological thinking. In that sense, it is not surprising that Robert Bellah, the coauthor of *Habits of the Heart* and the companion volume *The Good Society*, was deeply drawn to this discussion. The fact is that it was Bellah's teacher, Talcott Parsons, who stimulated attention among his students to the Axial Age after his own encounters with German philosophers and sociologists in the period before the Nazi seizure of power. Parsons's conception of sociology was notably idealist in character, however, and this set of emphases carried over to his students.

In what follows, I argue that, rather than only one Axial Age, there have in fact been three "Axial Ages," each with a distinctive character, which have decisively shaped human life since the birth of agriculture some ten thousand years ago. The "original" Axial Age was "moral" in orientation; the second, beginning around 1750, was "material" in character; and the third, contemporary Axial Age is chiefly "mental" in nature. The point of these designations is obviously not to suggest that they are the exclusive focus of human attention in these different periods, but rather that they set the dominant tone for each of those periods. It took more than two thousand years for the "material" Axial Age to follow the first, moral Axial Age. It did so because a particular concatenation of events having little (but not nothing) to do with the moral developments of the first Axial Age led to a breakthrough in human productivity unprecedented in human history (and largely inconceivable throughout most of that history). The idea of progress thus became a major preoccupation of writers such as Condorcet, because for the first time in human history it became a serious possibility.[2] The current mental Axial Age is, in turn, an outgrowth of the previous two, and it continues the efficiency orientation of the second Axial Age on an infinitesimal (atomic, molecular, or perhaps "nano") scale.

Our era will have to activate the moral and intellectual sensibilities cultivated in the first Axial Age if the human species is going to survive the costs generated by the extraordinary productivity gains of the second Axial Age. That is, the best traditions of modern science will have to combine with the most universalistic features of the Axial Age religions to pull the inhabitants of the planet back from the brink of self-destruction. There is every reason to think that intelligence and empathy can save us from the worst, so long as we use the tools we have been given by the Axial Age traditions to take seriously the challenges that confront us.

I would like to take this opportunity to thank Micah Kleit for encouraging me to put these ideas down in written form during a rather cursory conversation a couple of years ago about the state of discussion in sociology. I hope the result proves worthy of his enthusiasm. I am grateful to John Evans of UC San Diego, who provided comments on an earlier version in a panel at the Seattle meeting of the American Sociological Association, as well as to my old pal Bob Ratner, who went over the manuscript with his usual insightful and critical wit. I would also like to thank Emily Campbell, Danny Colligan, Lisa Krieg, and Austin Schmitz for their able research assistance. Danny generated the graphs that appear in the pages that follow; Austin put the manuscript into publishable form. Finally, I am grateful to the Graduate Center of the City University of New York for having provided me an intellectual home for the last decade. The Graduate Center is unique among institutions of higher education in the United States, and it is an important affirmation of the value of public higher education—an institution so crucial to our well-being today that, sadly, is losing the support of legislators and those who elect them. I hope this small book will suggest some of the worth the institution still holds.

The Three Axial Ages

Introduction

Much discussion has been devoted in recent years to the "Axial Age" described by Karl Jaspers and others, referring to the middle centuries of the first millennium BCE when Confucius, the Buddha, the Jewish prophets, and the Greek philosophers left their profound mark on human self-understanding.[1] In a post-World War II effort to bury the Hegelian philosophy of history that had privileged European development as the pinnacle of human achievement and to recognize the importance of other major world traditions, Jaspers described that epoch as having given birth to "man as we know him today" in the sense that certain ways of thinking emerged that created the moral universe in which we have lived ever since.[2] Objections to that conception have revolved around, among other things, the fact that the time period associated with the Axial Age was at once rather long—several centuries—and yet too brief to include such major developments in moral and religious thought as Christianity and Islam. This conceptualization led to various efforts to patch up the idea of the Axial Age with such notions as "secondary breakthroughs" or with the idea that there have been a number of "Axial Ages," each a significant "breakthrough" in human affairs.[3]

In what follows, I propose a sort of compromise position, asserting that there have in fact been three Axial Ages that have really mattered in human history: the "canonical," moral one that took place approximately twenty-five hundred years ago; what we might call a "material" Axial Age that began around 1750; and a "mental" Axial Age that is taking place today on the basis of the rapid improvements in information and communications technologies (ICT), artificial intelligence, robotics, and the like. The developments associated with these three periods form the fundamental matrix within which the human species currently finds itself and on which it must base its responses to the contemporary challenges facing it. This characterization of history in terms of three critical epochs seeks to make human history since the birth of cities recognizable and comprehensible for the student and the interested reader alike. I mean no disrespect to the practitioners of, say, the history of the Renaissance, who will surely insist that that era was decisive for understanding our contemporary predicament. But they know as well as anyone that that period did not "exist" before the Swiss historian Jacob Burckhardt breathed life into it in the mid-nineteenth century (and others have since identified other Renaissances in European history, particularly in the twelfth century). In contrast to Burckhardt, who stressed the emergence of individualism and other cultural developments, I find other features of the period since the decline of antiquity—and hence other dates—more significant. A brief characterization of the three epochs may help explain why.

First, each of the three Axial Ages has had a *characteristic preoccupation* and *attitude toward material goods*. The first, "moral" Axial Age, is associated with the idea of transcendence—the idea of another world against which this one could be measured and found wanting. It thus promoted ethical thinking and asceticism, and it is associated with the rise of several of the "world religions"[4] analyzed by Max Weber—Judaism, Buddhism, Confucianism—and with classical Greek philosophy. The second, "material" Axial Age began with what Reinhard Koselleck called the "*Sattelzeit*": the period beginning around 1750 when new technologies and British access to certain kinds of fuels (coal) combined to usher in astonishing and historically unparalleled advances in population growth, economic development, human health, and life expectancy. This "material" Axial Age created "more, more, more," with an attendant culture of hedonism. But it also created global economic inequality on a scale previously unknown in human history and, as would eventually become apparent, it did so at the price of the cannibalization of its natural substrate, widely referred to now as "the environment." Finally, the current "mental" Axial Age arises from the extraordinary breakthroughs in information, communications, artificial intelligence, robotics, and other technologies that are currently transforming everyday life.[5] The quintessentially mental tools of the third Axial Age are recasting human societies and making "virtual" and labor-saving technologies based on computers a central feature of human life. These new tools must be harnessed to the project of greater environmental sustainability if humankind is to rescue itself from the ecological damage caused by the methods of production associated with the second Axial Age.

Here it is worth noting that each of the three Axial Ages has relied on a different *energy regime*. The first Axial Age—like all of human history until the late eighteenth-century invention of the steam engine—rested on human and animal power. James Watt's innovation led the second Axial Age to depend on power generated from fossil fuels, which in turn made possible an unprecedented human-initiated transformation of the biosphere now increasingly referred to in geo-anthropological terms as the Anthropocene era. Finally, the third Axial Age—at least if the human species is to survive—will have to rely on renewable energy sources such as solar, wind, water, and other nonfossil fuels that do not generate heat-trapping emissions as do fossil fuels. As the world's population grows, gets wealthier, and lives longer, it will have to get more from less in order to feed, house, and maintain an aging and growing world population.

Finally, each of the three Axial Ages has had a distinctive *mode of thought*. The moral Axial Age, whose distinctive carrier groups were intellectuals and clerics, developed "thinking about thinking," as Robert Bellah has put it. That is, it witnessed a major advance in human self-reflection, partially because surpluses grew large enough to sustain people who could spend their time thinking rather than producing for their own survival. The second, material Axial Age was marked by "thinking about producing"; its pathbreaking thinkers were Adam Smith and Karl Marx, who both focused their attention on the creation and distribution of wealth in human societies. The third, mental Axial Age is distinguished from these earlier periods by "thinking that produces thinking." This quality is most apparent in regard to so-called artificial intelligence and "intelligent machines"; the mental Axial Age is the age of "smart"–smart phones, smart cars, smart houses, smart weapons, smart everything. It promises to provide the smart—that is, more efficient—technologies that may help make possible the rescue of the human species from its self-inflicted endangerment in the form of human-induced climate change. Yet it may also put many people out of work in the traditional, wage-earning sense, raising major questions about the relationship between work and income and about the value to the human species of these technologies more generally.

The third Axial Age has the task of redeeming the first two, applying the moral ideas from the first Axial Age, such as spiritual equality and harmony with nature, to the social and environmental problems generated by the second—which

TABLE 1

AXIAL AGE	CHARACTERISTIC PREOCCUPATION	ATTITUDE TOWARD MATERIAL GOODS	ENERGY REGIME	CHARACTERISTIC MODE OF THOUGHT
First (mid–1st millennium BCE)	Moral	Less is more (asceticism)	Human/animal	Thinking about thinking
Second (1750–1973)	Material	More, more, more (hedonism)	Fossil fuels	Thinking about producing
Third (1973–?)	Mental	Getting more from less (sustainability)	Renewable energy	Thinking that produces thinking

largely ignored these considerations in its phenomenally successful pursuit of "more, more, more." The second Axial Age brought about the advances in material well-being that Karl Marx once foresaw as the foundation of a truly human society, in which scarcity had been conquered and one could "hunt in the morning, fish in the afternoon, rear cattle in the evening, criticise after dinner . . . ," just as one might like.[6] But in the process of its massive increase in productivity, it turned the planet into . . . if not hell, a significantly warmer place than it had been before the Industrial Revolution. The year 2015 was the hottest ever recorded, and indications are that 2016 is on track to exceed that record (even if some of this is attributable to an unusually powerful El Niño, exacerbated by global warming). The third Axial Age will also have to manage the problem of insuring that the people whose jobs may become replaceable as a result of new technologies have a way to make a decent living. This concern has been around since the early nineteenth-century Luddites, of course, but it has taken on new urgency in the wake of recent developments in robotics and artificial intelligence whereby machines are increasingly making decisions once made by humans. Table 1 is a schematic representation of these essential features of the three Axial Ages.

The three Axial Ages have done a great deal to advance human well-being, yet they have left unresolved a variety of scourges that continue to beset the human species—class inequality, racism, remediable gender imbalances, ecological crisis, and violent death—even if a number of these have been remarkably improved by comparison with earlier times. Let me now explore each of the Axial Ages in turn.

1

The Moral Axial Age

It might be said that the first true "Axial Age" for humans was the transition from hunting and gathering to agriculture around 10,000 BCE. This shift in human patterns of survival has been associated with the onset of the Holocene geological period, which began with a warming of the earth that liberated many plant species from their frozen state in the last Ice Age.[1] The development of agriculture made possible the maintenance of larger human populations due to the creation of economic surpluses, but it also led to greater inequality among people. The inception of agriculture thus destroyed what the anthropologist Marshall Sahlins famously called "the original affluent society"[2] and nurtured differentiation in power and wealth as well as a deterioration of human health in certain respects.[3]

In due course, the spread of agriculture also led to the rise of cities—that is, "civilization" in the literal sense—around fifty-five hundred years ago. Civilization promoted refinement and luxury. With the emergence of cities, people no longer lived as Ibn Khaldūn described the nomadic inhabitants of the Arabian deserts, living lean off the land, well attuned to warfare, and capable of rapid movements across territory as fate demanded.[4] City-dwellers were sedentary, but they were often violent nonetheless—whether toward each other or toward their external enemies; much ancient literature and art were devoted to heroic tales of battle. But the early cities may well have been much safer than the Stone Age societies that agriculture replaced, where the rates of violent death may have been an order of magnitude greater than those registered in the twentieth century, so fabled for its sanguinary history.[5] One must keep this earlier background in mind when considering subsequent developments.

The notion of the "Axial Age" associated with the German philosopher Karl Jaspers centered on its intellectual and moral consequences and had little to do with other aspects of the period. Jaspers focused on the religio-intellectual developments in China, India, ancient Palestine, and Greece, arguing that they shared a new stance toward the world. Confucius, Siddartha Gautama (the Buddha, or "Enlightened One"), the Jewish prophets, and the Greek philosophers were said to have contributed a new skepticism toward this world and a standpoint from which to critique it. According to S. N. Eisenstadt, one of Jaspers's most influential followers, the Axial Age invented the idea of "transcendence": "the perception of a sharp disjunction between the mundane and transmundane worlds . . . [with] a concomitant stress on the existence of a higher transcendental moral or metaphysical order which is beyond any given this- or other-worldly reality."[6]

Despite Eisenstadt's emphasis on the transcendental, however, the intellectual breakthroughs associated with the Axial Age were by no means all "religious" in the narrow sense. Indeed, as Max Weber pointed out, Socrates' "ethical and strongly utilitarian rationalism . . . cannot be compared at all to the conscience of a genuine religious ethic; much less can it be regarded as the instrument of prophecy."[7] Similarly, a scholar of Chinese civilization has recently insisted that "imperial China was the only world civilization where transcendental world religions exerted no major influence on politics."[8] Accordingly, Johann Arnason has noted that "it has proved difficult to subsume the whole spectrum of the Axial Age under an idea of transcendence a la Jaspers; critics have also noted problems with Eisenstadt's broader conception of 'transcendental visions.'"[9] Notwithstanding these objections to the centrality of "transcendence" to the first Axial Age, the era is generally conceived as a phenomenon associated chiefly with heightened attention to morality and self-reflection facilitated by a conception of other and better worlds.

In part to accommodate the importance of the less obviously "religious" developments, Robert Bellah's approach emphasized that "second-order thinking," or "thinking about thinking," is decisive for understanding the Axial Age.[10] Bellah quotes at some length a passage from the historian of antiquity Arnaldo Momigliano as making the point more generally: "New models of reality either mystically or prophetically or rationally apprehended, are propounded as a criticism of, and an alternative to, the prevailing models. We are in the age of criticism."[11] Bellah and Momigliano thus stress the critical stance toward existing reality that they regard as typical of the period. Similarly, Nicholas Baumard and his colleagues have described the first Axial Age as a time when "ascetic wisdoms and moralizing religions" came into being for the first time, moving religion beyond its previous ambit of "performing rituals, offering sacrifices, and respecting taboos in order to ward off misfortune and ensure prosperity."[12] All this highlights the extent to which the first "Axial Age" has been conceived in intellectual-philosophical-moral—but not necessarily religious—terms, and largely ignored other considerations of the period. There can be little doubt that the epoch did, indeed, witness a major transformation in human self-understanding, at least in those parts of the Eurasian *ecumene* that experienced them—from the eastern Mediterranean to the Middle Kingdom. Gradually, these developments would spread throughout Eurasia and beyond, and as the Axial Age ideas and their descendants (especially Christianity and Islam) assumed the role of ideological cloaks for conquest and domination, they would overwhelm many practitioners of "indigenous" or "traditional" religiosities in areas outside the original domains of the world religions.

Although the idea of the Axial Age is generally attributed to Jaspers, Eugene Halton has recently dissented from the view of the Axial Age that ascribes it to Karl Jaspers and then, following him, to Talcott Parsons's protégés, among them Shmuel Eisenstadt and Robert Bellah.[13] Jaspers himself had already noted that something like the idea of the Axial Age had previously been advanced by such authors as Abraham Hyacinthe Anquetil-Duperron and Ernst von Lasaulx.[14] For his part, Halton attributes the idea chiefly to a now-forgotten Scottish folklorist named John Stuart-Glennie, who developed his theory that a "moral revolution" had taken place in the world in approximately the sixth century BCE and presented it, among others, to the Sociological Society of London in 1905. However, Stuart-Glennie's approach was more holistic and less purely intellectualistic than that of Jaspers and colleagues. Halton takes great interest in the Scotsman's insistence on what he called "panzoonism" as the foundation of an adequate view of the human situation. According to Halton, panzoonism—a mode of thinking "from a life-based perspective" rooted in the so-called bioticon of which humans were a part—is a stance more sympathetic to the putatively more holistic and eco-friendly outlooks of aboriginal peoples.[15] In other words, Stuart-Glennie's understanding of the "moral revolution" of the sixth century was more open to the importance of faith traditions other than those of Weber's "world religions" and, in particular, to those that seem to offer a more promising perspective in the face of contemporary climate and ecological crisis.

Leaving aside the question of its progenitors, what *caused* the first Axial Age? Some have argued that the intellectual-moral developments of the period were associated with "breakdowns" in the social and political order. Though he never used the term "Axial Age," for example, Max Weber did speak of a more or less simultaneous "age of prophecy" and suggested that "perhaps prophecy in all its forms arose, especially in the Near East, in connection with the reconstitution of the great world empires in Asia."[16] Max Weber's brother Alfred, commenting on the emergence of a "synchronistic world age" during this period, attributed it to the "great migrations" of the early first millennium BCE.[17] Jaspers himself spoke of "common sociological preconditions" of the Axial Age developments in the different regions of Eurasia, such as social upheaval, wars, and revolutions.[18] For Robert Bellah, however, there have been many "breakdowns" without breakthroughs, and hence he tended to dismiss such events as primary causes of the breakthroughs.[19] In contrast to this "breakdownist" approach to the matter, recent scholarship has suggested that the Axial Age may actually have been a by-product of ancient prosperity—or at least of such limited prosperity as was available in an agricultural society.

From this perspective, the characteristic preoccupations and modes of thought of the first Axial Age were a product of an early version of what the political scientist Ronald Inglehart, trying to make sense of the rise of ecological and other kinds of political issues in recent decades, has called "post-materialism": according to Baumard and his coauthors, "increased affluence explains the emergence" of the "ascetic wisdoms and moralizing religions" of the first Axial

Age.[20] Their more empirically grounded explanation nonetheless fits with that of Eisenstadt, who stressed above all the importance of the appearance of a sociological stratum—intellectuals and the clergy—that had the freedom from necessity and the intellectual tools with which to develop these ideas.[21] Yet while Eisenstadt may correctly stress the importance of the "rise of clerics" as a group with the freedom to develop ("rationalize") these ideas, that is different from explaining the shift toward specifically ascetic and ethical doctrines. Baumard and colleagues argue that the turn toward asceticism and moralism can be accounted for by "life history theory," which distinguishes between "fast" strategies emphasizing the pursuit of short-term goals and "slow" ones that promote long-term investments—as, for instance, in reciprocity rather than self-seeking. In other words, a rise in affluence could reassure people that they need not focus so intently on short-term advantages, thus strengthening impulses toward generosity, altruism, and forbearance.[22] Yet Baumard and colleagues also note that the size of their sample of ancient civilizations that might have produced Axial Age religions is very small, that their measures of affluence are rudimentary, and, above all, that their findings cannot account for the different "directions" of these religions' "rejections of the world," to borrow Max Weber's characterization of their stances. Meanwhile, others have argued that the moralization of religion took place earlier in some places, later in others, and that other factors besides greater economic well-being may have been decisive.[23]

Even if Baumard and colleagues provide a reasonable explanation for why asceticism came to the fore and religion came to be "moralized," they cannot account for the different specific "directions" taken by these varied doctrines. For that task, Weber's emphasis on these doctrines' various "carrier groups" and their predilections is essential. In an attempt to summarize them "succinctly, in a formula, so to speak," he wrote, "In Confucianism, the world-organizing bureaucrat; in Hinduism, the world-ordering magician; in Buddhism, the mendicant monk wandering through the world; in Islam, the warrior seeking to conquer the world; in Judaism, the wandering trader; and in Christianity, the itinerant journeyman. To be sure, all these types must not be taken as exponents of their own occupational or material class interests, but rather as the ideological carriers of the kind of ethical or salvation doctrine which rather readily conformed to their social position."[24] As in so many other areas, Weber deployed ideas first developed by Marx—in this case, that people's ideas tend to be a product of the way they live their lives—but often applied them more consistently than had Marx himself.[25] The specific orientations assumed by the moralizing wisdoms and religions spawned during the first Axial Age must be explained in considerable part in terms of the inclinations of those who first developed them, even if they would subsequently undergo further development.

Weber makes a crucial distinction between two forms of prophecy reflected in the world religions spawned during what he called the "prophetic age" (reaching from the eighth century BCE until as late as the fifth).[26] The East and South Asian forms of prophecy, for which the Buddha set the standard, were said to be "exemplary" in nature, whereas those originating in the Near East and typified by Zoroaster and (much later) Muhammad were characterized as "ethical" in character. "Exemplary" prophets enacted a form of behavior that was to be emulated by the followers, who were expected to become "vessels of the divine"; "ethical" prophets enjoined certain commandments, and adherents were to become "tools of the divine." The differences between these two basic types of prophecy, according to Weber, set the respective world regions on divergent paths that would lead to profoundly different religio-cultural trajectories—and eventually, of course, divergent economic and social outcomes as well.

But then, beyond the ethical/exemplary distinction, there were also significant differences in emphasis and outlook among the various religions that emerged across the Eurasian landmass over the past three millennia as well. Let us begin with India. The so-called Aryan invaders who entered India from the northwest in roughly the middle of the second millennium BCE were nature-worshippers. That archaic form of religion would, in time, be superseded by Vedic Brahmanism—a priestly religion associated with "books of knowledge" (*vedas*) that were also poems addressed to the gods seeking their approbation and assistance. The priestly stratum inherited from the ancient poets the responsibility for purveying, whether by writing or in speech, the sacred hymns and mantras, which were regarded as powerful magic and accessible only to the religiously qualified.

The canonical text of Brahmanism was the Rig Veda, which is thought to have been composed well before 1000 BCE but not written down before about 600 BCE. Like the Qur'an, the Rig Veda was thought to have been divinely revealed poetry and was followed by prose commentaries that elaborated the rules and rituals of the faith. Because they marked out a special role for the Aryan priestly class, the commentaries were referred to as the Brahmanas. Vedic religion was organized to a high degree by sacrifice, and as in Zoroastrianism, the dominant religion of ancient Persia, sacrifice was unfailingly accompanied by fire. Indeed, the god of fire, Agni, was the most frequently mentioned god in the Rig Veda. Aryan households were expected to offer sacrifices to the gods five times daily, either by the Brahman head of household or by his Brahman priest or cook, from whom anyone would take food because of his high, nonpolluting status.[27] As guardians of and go-betweens with the spiritual world, the Brahman priests had come by the middle of the first millennium BCE to occupy positions of earthly power as well as of religious authority.

In its search for an explanation of the cosmos, the Rig Veda outlines a social hierarchy that sprang from the anatomy of the original cosmic man: the *brahmans*, or priests, came from his mouth; the *kshatriyas*, the warriors, from his arms; the *vaishyas*, the merchants, artisans, and farmers, from his legs; and the *shudras*, the servants, from his feet. The *shudras* were probably originally the indigenous people conquered by the lighter-skinned Aryans from western Asia, and the emergence of this category to supplement the three-class order familiar from European society (those who pray, those who fight, and those who work) helped institute a profound color-consciousness that ever since has been "a significant factor in reinforcing the hierarchical social attitudes that are so deeply embedded in Indian civilization."[28] These social attitudes characterized Indian society until very recent times, although caste and related forms of hierarchy were outlawed under the Indian constitution of 1949.[29]

When in the middle of the first millennium the Magadha dynasty came to dominate a substantial portion of the Ganges plain, the Brahmans had come, along with the warrior class (*kshatriyas*), to be crucial pillars of rule, though they were also much less political in their self-understanding than Confucian sages. Instead, they were becoming an educated status group, owning land and controlling village life through their dominance of ritual. Meanwhile, the quadripartite *varna* system had become a pervasive element of northern Indian social life. The *shudras*, who brought up the bottom of the social and religious hierarchy, were regarded as ritually unclean and hence consigned to the lowest occupations. (Eventually, another group, the absolutely unclean "untouchables," would be added as well.)[30] More especially, they were forbidden to hear the Vedic mantras, possibly on pain of death or excruciating punishment, such as having molten lead poured in their ears. By this time Brahmanism had become associated with a traditionalistic defense of the privileges of the Brahmans and of ritual—a ritual that was strongly bound up with sacrifice and hence with killing. Of course there was a good deal of killing going on in the society at large, despite the Magadha conquest of the Gangetic plain. The *Mahabharata*, the epic tale thought to reflect Indian life in approximately 1000 BCE but composed several centuries later, is "drenched in the blood of endless warfare."[31] The Brahmana era was thus one in which social hierarchy had become sharp, political authority centralized and unprecedentedly powerful, and religious life to a considerable extent a defense of this power and inequality.

In this context, there arose disquiet among thoughtful people concerning the direction of their society, provoking a variety of novel responses to their situation. Perhaps the first development to discuss is the Upanishads, the first of which were composed in approximately the eighth century BCE but whose canonization was not secured until the late centuries of the first millennium BCE. In contrast to the stress in the Rig Veda on mantras and sacrifices as the means to salvation, the Upanishads promoted the idea of wandering in the forest in search of understanding. Thus emerged the figure of the mendicant ascetic who renounced conventional life in favor of the pursuit of wisdom; in the process, a rationalistic side was imparted to Indian religiosity. Along with the quest for wisdom, the Upanishads instructed that yoga exercises could prepare the body for the abandonment of human striving, desire, and their consequent frustrations. The "quiescent, mindless, motionless inactivity of *moksha*"[32]—"release" from suffering—now became the chief goal of Vedantic meditation. *Moksha* could be achieved by knowledge of the self, which required control over oneself and one's desires. At the same time, the idea of samsara—an endless cycle of birth, death, and rebirth from which the individual can be freed only with great difficulty or good fortune—comes to the fore. The linkage between karma and samsara emerges here as "a distinguishing axiom of Indic civilization."[33]

Max Weber viewed the idea of karma as the most complete, systematic solution of the "theodicy problem"—the question of the relation of god to the world—ever invented. Weber shows that the idea of karma entails the creation of a world in which all acts have consequences for the person's fate in the next life; the person is enjoined to behave well in order to merit rebirth in a better station or, ideally, to be released from the wheel of rebirths entirely. As a result of this all-encompassing conception of the place of ethical action in the world, "the dualism of a sacred, omnipotent, and majestic god confronting the ethical inadequacy of all his creatures is altogether lacking."[34] Weber appears to have been contrasting the karma doctrine and its consequences with the ethical implications of the Jewish god created during roughly the same period in Israel/Palestine, which did create such an omnipotent god. Weber distinguishes between "ethical" and "exemplary" forms of salvation religion, and Hinduism clearly lies on the latter side of the divide between these two types. Robert Bellah seems to go even further in his evaluation of the basic orientations of Hinduism. He suggests that the durability in modern Hinduism of the ideas of the Rig Veda, which reflect the attitudes of a pre-state society, may entail the persistence in Indic religiosity of a tribal sensibility that "raises questions about the whole idea of religious evolution,"[35] which his magnum opus seeks to defend. Bellah seems to be suggesting that, from the standpoint of other world religions, there is something uniquely archaic or atavistic about the fundamental impulses of Indic religiosity. Indeed, Hinduism's deep implication in the caste system seems hard to dispute—especially if one considers the extent to which Hindus have been prepared at various times to abandon the fold for more egalitarian doctrines such as Islam and Christianity.

While the concepts of karma and samsara would play a crucial role in subsequent Indian religion, this did not necessarily mean that they always played the same role in different religious traditions. Consider Jainism, which arose around 500 BCE and persists today as a small minority religion in India, perhaps comparable to Mormonism in the United States in its distinctness from but rootedness in the tradition from which it emerged. At the outset, however, Jainism was a major source of a new worldview. The religion was associated with the teachings of Mahavira, a northern Indian aristocrat who renounced his privileges and adopted a peripatetic life of teaching and contemplation. The Jains argued that all things, living or nonliving, had a soul, and they viewed karma—the principle of causation—as a substance that accumulated upon the soul to the extent that the individual acted in the world. Thus the Jains understood salvation in terms of "the escape of the soul from its adhering karma and its upward flight to live on forever in a realm of pure bliss."[36] Salvation could only be achieved by adherence to an extremely stringent series of rules that forbade killing, stealing, lying, sexual activity, and the ownership of property. The prohibition on killing and on causing any suffering—which the West came to know, especially through Mahatma Gandhi, as the principle of ahimsa—was interpreted so strictly that Jains could not kill their own food; it had to be harvested by laymen. Indeed, Mahavira is thought to have slowly starved himself to death in his attempt to live up to the strictures against killing. To this day, Jains will sweep off a chair before sitting down so as to insure that no living thing might die in the process. Due to the prohibition on harming living things, devout Jains have tended to avoid agricultural occupations and to be urbanites in mercantile occupations.

The rewards of such a life of voluntary privation might seem paltry to rich Westerners, but it continues to attract adherents—although understandably relatively few. A thirty-eight-year-old Jain nun recently had this to say about the rigors of her faith:

> People think of our life as harsh, and of course in many ways it is. But going into the unknown world and confronting it without a single rupee in our pockets means that differences between rich and poor, educated and illiterate, all vanish, and a common humanity emerges. . . . This wandering life, with no material possessions, unlocks our souls. There is a wonderful sense of lightness, living each day as it comes, with no sense of ownership, no weight, no burden. Journey and destination became one.[37]

Like Mahavira, this Jain ascetic had resolved to commit suicide by starvation, the ultimate act of selflessness deriving from the Jains' understanding of ahimsa. Perhaps necessarily, at the time that it emerged, Jainism represented an extreme version of the ascetic renunciation that would characterize much subsequent Indian religiosity.

Buddhism arose in this same historical context; Mahavira and the Buddha were roughly contemporaries, living on either side of the year 500 BCE. Siddhartha Gautama was a young noble who could look forward to a life of power and comfort, but he threw it all away in favor of wandering in search of the truth of existence. That truth, he concluded, was that this world was essentially an illusion, and that there was little to seek in this life. In contrast to Jainism, there were no souls or selves to save in Buddhism. Instead, the aim was to avoid accumulating karma by stanching desire. Despite the lack of a soul to which karma could adhere, as in Jainism, Buddhists nonetheless regarded the accumulation of karma as inevitable for the unenlightened. They thus sought release—*moksha*—from entanglement in the world and the attainment of nirvana, a state beyond desire and hence beyond its frustrations. This exalted state was attainable more or less only by ascetic monks, who relied on laypeople for alms and donations. Those laypersons were at best likely to be able, through proper performance of the duties of their station, to be re-born as beings somewhat closer to enlightenment, release, and *nirvana*. In these respects, while diverging from the religious practices and ideas of Vedic religion and lacking any stake in its social arrangements, Buddhism failed to challenge very much the social structures inherited from Brahmanism.

The Buddhist *sangha* (brotherhood) gathered in monasteries, which gave them an organizational backbone and, in time, a thriving economic base. The monasteries symbolized the Buddhists' retreat from the world, even if they did not entirely abandon workaday life because they depended on the laity for their upkeep. Having started as an effort to purify Brahmanism and rid it of its objectionable qualities from *within* educated Brahman circles, Buddhism gradually came to challenge it from a separate and organizationally superior position. In the meantime, Buddhism spread "by filling the moral vacuum in the new social world of commerce and city life with a universalistic social morality" absent from Brahmanism.[38]

Taken together, the emergence of Jainism and Buddhism might be seen as Vedic Brahmanism's Protestant Reformation. Early Buddhism resembles Lutheranism in the sense that it rebelled against the worldliness of its "mother ship"—in this case, Brahmanism—finding in the more established tradition a distraction from piety and a misuse of faith, and it offered a refurbished doctrine meant to re-secure the connection to the sacred under very different terms—especially the doctrine of nonviolence (ahimsa). Buddhism also was marked by a greater "reasonableness" than Jainism, a more relaxed quality comparable to Lutheranism's less extreme rejection of established ecclesiastical authority when compared with that of the Reformed Protestant sects. Jainism was in this sense more similar to Reformed Protestantism, which was marked by rather greater rigor in its approach to keeping the faith. Like Calvinism, Jainism took for granted that the greater the suffering of the believer, the better for his or her soul. Meanwhile, Buddhism was the only movement of its time to require that its monks preach to the laity, maintaining important links to them when others such as the Jains insisted on pure virtuoso religiosity. Finally, however, having conceived of the world as chiefly a realm of suffering, Buddhism's ethics were somewhat more oriented toward relieving the suffering of others than Lutheranism's insistence on "faith alone" as the guarantor of salvation. Since Buddhism valued "good karma," it was also prepared to countenance good works as appropriate efforts on the path to salvation. Jainism encouraged non-harming as a path to salvation, but commanded little else in the way of helping others. In that regard, it might be said to have promoted a more individualistic orientation to the world, one not especially concerned about the fate of others. In this it differed notably from Reformed Protestantism's stress on a God-fearing community intently concerned about the fate of one another's souls.[39]

The "Protestant Reformation" analogy goes only so far, however. In contrast to the Christian faiths, which remain separated despite centuries of ecumenical soul-searching, Hinduism in India later would reinterpret the Buddha as a reincarnation of Lord Vishnu and reabsorb Buddhism into itself. Like a stream that divides in one place, only to converge again further downhill, Hinduism first spawned and then ingested Buddhism, such that it largely disappeared from its Indian birthplace, with the final coup de grace being administered by hostile Muslims in the early second millennium CE. Carried off by wandering monks seeking a more hospitable place to worship, Buddhism subsequently came to be more associated with East and Southeast Asia than with the land of its historical origins. Buddhism and Jainism emerged at first from the Brahmanic context, but their rejection of some of its elements and elaboration of novel emphases led to the foundation of new faiths with common roots but also with very different features.

Next, we must examine a more or less contemporaneous development on Asia's western fringe, in what we now call the Middle East. This was the emergence of prophetic Judaism, which invented a radically new relationship between a people and their god(s), as intimated above in Weber's discussion of the karma doctrine. We should first of all clarify the meaning here of "prophecy." This activity had less to do with our sense of "divining the future" and more to do with the interpretation of the will of the gods. Thus, although they were sometimes foretelling the downfall of some tyrant for misbehavior, the Jewish prophets were chiefly articulating their understanding of the gods' commands to the people. In doing so, they reinforced the idea that the Israelites should follow one god above all others. This stress on the worship of one god to the exclusion of others, or at least above those others, arose against the background of intense pressure on the Children of Israel, those who believed that their god had given them special claim to lands at the eastern end of the Mediterranean. The principal early kings of ancient Israel, Saul, David, and Solomon, sought to erect a strong state that would protect the Jews from their antagonists, especially the Assyrians, Egyptians, and Philistines. Yet Solomon's state-building efforts made so many demands on the populace that, after his death in c. 925 BCE, resistance arose and the kingdom became divided into a northern part and a southern part—Israel and Judah, respectively. It is from the latter that we have our name for the people and their religion. This period of the divided kingdom "corresponds roughly to the beginning of the continuous development and historical importance of the Hebrew practice of prophecy."[40]

The Jewish prophets of the early first millennium BCE are traditionally thought to have built on the cult of Yahweh. Recent scholarship, however, suggests that the original god of Israel was not Yahweh, but El, and that ancestor worship was perhaps more important in ancient Israel than later became the case. The importance of El rather than Yahweh in early Jewish history helps make sense of the very name "Israel" (*Yisra-el*), which literally means "he who strives with god."[41] Thus was created a conception of god as one with whom a people wrestles, at least metaphorically. In the words of historian of Christianity Diarmaid MacCulloch, the Jews are unusual in that they "struggle against the one whom they worship. . . . The relationship of God with Israel is intense, personal, conflicted."[42] That relationship was also novel in the sense that it was based on a covenant binding the people to their god and creating a powerful bond between them.

According to Robert Bellah, this view of the relationship was borrowed from Assyrian precursors, which however involved king and subjects, not god and people. In approximately 722 BCE, the Jews in Israel suffered destruction at the hands of the conquering Assyrians. The Assyrian ruler, Aššur, demanded loyalty to himself as god-king. As victims of the Assyrian onslaught, the Jews rejected those demands in favor of fealty to one god and one god only—Yahweh.

The name of the ninth-century prophet Elijah, which means "Yahweh is my god,"[43] suggests the growing convergence of El with Yahweh and their gradual acceptance as synonyms. Insistence on the exclusive worship of Yahweh arose as an effort by the prophets, who were keenly attuned to the external threats to the Hebrews, to promote one particular god as the savior of the Jewish people. The choice of Yahweh was not foreordained. The period of the divided kingdom had "produced kings prepared to experiment with the gods of more powerful people" as possible protectors.[44] The fertility god Baal of the old Canaanite pantheon thus became one of the chief gods against which the prophets inveighed in their endorsement of Yahweh as the principal god of the Hebrews.

In the process, the prophets created the notion of a remote, all-powerful creator deity to whom they owed their chief obedience, above and beyond that to any earthly being. "A God who is finally outside society and the world provides the point of reference from which all existing presuppositions can be questioned," writes Bellah. "It is as if Israel took the most fundamental symbolism of the great archaic civilizations—God, king and people—and pushed it to the breaking point where something dramatically new came into the world."[45] That something new was the idea of a people beholden to their obligations to their god, and vice versa. Against this background, the Hebrew prophets called the Jewish people—individually and collectively—to live up to their covenant with Yahweh/God. This emphasis frequently brought them into conflict with the earthly rulers of the people. For example, the very idea of a monarchy could be seen as inconsistent with the obligation to give one's obedience to god, and thus there remain to this day orthodox Jews who object to the existence of the state of Israel as a blasphemous betrayal of Jewish ideals.

Then there were the criticisms of the specific policies of the kings. Thus the prophet Elijah inveighed against the corruption of King Ahab and his wife Jezebel and prophesied his doom. In the eighth century BCE, the Judean prophet Isaiah chastised the wealthy for their exploitation of the poor. But ultimately the message of the prophets was that it was the duty of every individual to honor the covenant with Yahweh. The ethical duties were outlined in the Ten Commandments, traditionally ascribed to Moses's divine inspiration. Obedience to God's commandments would, in turn, result in God's approbation and the security and prosperity of the people. The notion that worldly success was a reward for religiously correct behavior promoted a non-ascetic orientation to the world. As prophetic Judaism developed, however, there emerged a deep and abiding tension between the particularism of the Jews' covenant with Yahweh and the implicit universalism of the ethical creed. Yahweh's commands were, in principle, valid for everyone, due to his remoteness and omnipotence. This all entailed a powerful push toward ethical concerns in Judaic religion and its successors.

As is well known, Judaism was the progenitor of Christianity, but the latter's conception was not, so to speak, immaculate; instead, Christianity was the product of a "Hellenized" Jewish culture that took root on the eastern fringes of the Roman Empire. Its outlook derived in many ways from Greek precursors, filtered through Roman re-workings of that inheritance. In order to make sense of Christianity, we must therefore consider some of the elements of classical Greek culture that echoed down into later developments.

The Greeks of fifth-century Athens famously created a form of social organization widely regarded as unprecedented among state-bearing societies. That organization was the *polis* or "city-state," the institution from which we derive our words for "politics" and "policy," not to mention "police." While it is often thought that the form of government in the fifth-century *polis* was "democracy," this was by no means necessarily the case; *poleis* could be ruled by illegitimate usurpers, oligarchies, or the mob. Even when the *polis* was ruled "democratically," we should understand that this democracy involved only about one-fifth of the adult male population. Among those who participated, however, it was a *direct* democracy, relying on face-to-face assemblies and debates over matters concerning the city-state. The Athenians thus invented a form of government that included relatively many ordinary people and that, more importantly for our purposes, made decisions on the basis of persuasion and without divine sanction. The link between divinity and rule, so characteristic a feature of previous states, was broken. Even if the gods were very much part of their world, the *ecclesia* ("assembly," but later the basis for the Italian *chiesa* [church] and of our word *ecclesiastical*) that made decisions for the *polis* did not need the sanction of the gods to do as they chose to do.

This human-centered mode of decision-making was strongly reflected in classical Greek art and sculpture, which was obsessed with representations of the human form. This enthusiasm for the human body went so far as to insist that athletic games be conducted in the nude, a practice that the Romans and others would later find positively embarrassing. To be sure, the emphasis was on the male form above all, and classical Greece certainly remained a man's world. Even male homosexuality was exalted above other forms as the most fitting expression of human love.

Greek religion was also remarkably human-centered for its time. In contrast to the phantasmagorical figures we associate with Indic religiosity, the Greek gods were represented as larger-than-life versions of human beings and conceived very much in human terms. Temples were homes for the gods, not places of worship. Meanwhile, in classical Greece of the fifth century, there was no separate priestly stratum; Greek priests were officials of the state "in exactly the same sense as generals or treasurers or market commissioners . . . with the same tenure and rotation of office, as the others."[46]

Meanwhile, the bards who passed down the stories from the heroic period that followed the collapse of Mycenaean civilization around 1200 BC were just that—poets—not priests on the Brahmanic model who guarded sacred knowledge and dispensed it for a fee. The gods played a central role in those stories, and according to Herodotus, it was Homer and Hesiod who "'first fixed for the Greeks the genealogy of the gods, gave the gods their titles, divided among them their honours and functions, and defined their images.'"[47] But, unlike the Vedas, which were the property of the Brahmana priests, the stories were understood as the common possession of all Greeks, and while the gods were very humanlike, the degree to which human beings in those stories were also raised up to a level close to the gods was striking.

The Greeks also exhibited a pronounced desire to make sense of the empirical world, a practice that we now know as philosophy. Two of the most remarkable of the "lovers of wisdom," Plato and Aristotle, bequeathed to posterity resources for thinking about the world that continue to generate insight and discussion. For present purposes, Plato is perhaps the more important figure, as he contributed ideas about religion that would have profound ramifications later on. First, he advanced a conception of god as perfect oneness. Like the Jews, Plato also stressed transcendence in his view of god. Yet, in contrast to the god of the Hebrews, who were busy wrestling with each other, Plato's conception of a perfect god was passionless and unchanging. This would later help provide Christianity with a conception of divinity as remote and impervious to influence, the "*deus abscondidus*" that Weber argued left the Calvinist in "an unprecedented state of inner loneliness"[48] about the fate of his or her soul. In addition, Plato mused about the existence of an individual soul, a spark of immortality that outlived and was ultimately more real than the mortal housing in which it existed in this life. Such a view differed sharply from the Jews' attitudes on the matter at this time; until the time of the Maccabean revolt in the second century BCE, Jewish writings on the matter suggest "that human life comes to an end and, for all but a few exceptional people, that is it."[49] The preoccupation with life after death would prove one of the major features that would permit later observers to regard "Christianity" as something other than the Judaism from which it had originally emerged.

In some ways like the Greeks, the Chinese during the middle of the first millennium BCE also underwent a period of profound reflection on the human being's place in the world. Just as Plato has been the central figure in Western philosophy since he taught, Confucius has played a similarly decisive role in subsequent Chinese and East Asian thought—to the point that Chinese regimes seek to invoke or suppress his teachings in accordance with their calculations regarding their prospects for maintaining power. (The government of Xi Jinping currently seeks to venerate Confucius as a source of social stability in times of economic and social upheaval; the Chinese have been creating Confucius Institutes around the world to promote the country's image in international circles.) Yet the teachings of Confucius revolved around questions of correct ritual and of goodness rather than of the individual's relationship with one or more divinities. Hence, like the developments associated with classical Athens, Confucianism has long been regarded as strongly "secular" in character. Against the background of much violence, Confucius was above all concerned with recovering the supposed harmony of earlier days by way of a return to the proper performance of ritual. Confucius insisted that, while it was important to be in tune with the desires of Heaven, human beings choose their own fate. They could thus choose to do right or wrong, and the cultivation of "the gentleman" involved training toward inner reserve and self-control.

The "secular" qualities of Confucianism have long raised questions about whether these doctrines should be regarded as a "religion" or not. For Weber, "Confucianism is rationalist to such far-going extent that it stands at the extreme boundary of what one might possibly call a 'religious' ethic."[50] In *The Religion of China*, Weber describes Confucianism as a philosophy of rational adjustment to the world, but one that could also coexist peaceably with the "magic garden" of Taoism.[51] Indeed, one of the remarkable features of Chinese religiosity is its capacity for cohabitation among a variety of belief systems. Chinese religion thus lacks the tendency toward exclusivism that characterizes all the major faiths that originated to its west—Judaism, Christianity, Islam, and Hinduism. This was surely because the Chinese state "was more willing [than other major empires of antiquity] to adopt a pluralistic attitude toward religions, and . . . countless popular religions not only flourished alongside the world religions, but also sometimes developed at the expense of the latter."[52]

In any case, Weber contrasted Confucianism's accommodative impulses toward the world with the otherwise similarly rationalistic impulses of Puritanism. In Weber's view, whereas Confucianism sought to inculcate adjustment to the world as it is, Puritanism instilled the sense of a tension with the world that led the Calvinist to want to transform it. It made the Puritan into a "tool of the divine"; by contrast, the Confucian gentleman could not be anyone's "tool." "The Confucian owed nothing to a supra-mundane

God; therefore, he was never bound to a sacred 'cause' or an 'idea'"; instead, "the duties of a Chinese Confucian always consisted of piety toward concrete people whether living or dead. . . ."[53] In sum, according to Weber, Confucianism inculcated a radically "optimistic" outlook that was at odds with the dictates of an ethical god enjoining action in this world to make it more consistent with that god's strictures about how the world should be. It was thus ethically more "this-worldly" than reformed Protestantism and the latter's insistence that the believer make the world into an ethically livable place.

In his extraordinary study of *Religion in Human Evolution*, Robert Bellah has argued recently that this understanding of Confucianism is fundamentally wrong-headed. In Bellah's view, Confucius's insistence that the rulers properly observe the time-honored rituals supplied the foundation for the notion that China's rulers needed to serve the interests of their subjects. Otherwise, their legitimacy as rulers would be squandered. This is the background to the idea that Chinese rulers have "the mandate of heaven," but also that they can lose that mandate if they fail to live up to its requirements. Beyond this, however, there is also in Confucianism "an ideal of human self-cultivation leading to an identification with an ultimate moral order, with the *Dao* and the will of Heaven, that is available to individuals, however grim the social situation and however much they may seem to have 'failed' to bring good order to society."[54] Confucianism and other Chinese systems of thought thus instilled in both the Chinese and their rulers moral standards that entailed a tension with the prevailing order of things; pace Weber, these systems of ideas were not merely modes of adjustment to the status quo. Thus, Bellah argues, "given Weber's enormously influential analysis of China as a stagnant, traditional society, it is perhaps well to point out that such was not the heritage of the Axial Age to later Chinese history."[55] The reason for Weber's misunderstanding may perhaps be traced to his preoccupation with the "economic ethics of the world religions" and the reasons that non-Western civilizations failed to "take off" economically, rather than with their political proclivities. Yet, as we will see below, recent scholarship also suggests that Weber overstated the extent to which religious ideas were decisive in the "great divergence" of the West from the rest.

To sum up: Reflecting its heavily Weberian inflection, the first, "moral" Axial Age as propagated by Jaspers and his followers has been understood almost exclusively as an intellectual and moral transformation, and one concentrated in the "Old World" of the Eurasian ecumene. Jaspers's notion of an Axial Age—the "canonical" one thus far, notwithstanding Halton's criticisms—privileged the historic centers of Eurasian high culture at the expense of the cultures and religious worldviews of those elsewhere and later, even though some way had to be found to include Christianity and Islam as part of the modern outlook. Meanwhile, the "indigenous" religions of Africa and the Americas are frequently said to articulate or envision a harmonious relationship between humans and nature and hence to be more relevant to our current situation than other (for example, "Western") ways of thinking about the world. Yet these "indigenous" worldviews are left out of the conception of the Axial Age, suggesting a potential undervaluing of traditions and worldviews that did not spread along with a successfully conquering army. More critically, perhaps, some historians reject the entire Axial Age enterprise as an unwarranted intrusion of philosophers and theologians into the domain of the historian. According to the prominent German Egyptologist Jan Assmann, "The idea of the Axial Age is not so much about 'man as we know him' and his/her first appearance in time, but about 'man as we want him' and the utopian goal of a universal civilized community."[56] For his part, Halton's concern is that no universal civilized community could ever ensue from the religious traditions whose origins are stressed in the debate over the Axial Age. Like many in the contemporary world, he believes we must make a much more radical transition in our thinking if we are to avert the ecological consequences of the Second Axial Age. If we are to avert the worst, the moral resources of the first Axial Age are going to be crucial resources on which to draw.

2
The Material Axial Age

The first Axial Age was distinct from what had gone before chiefly in intellectual terms, a fact whose significance is heightened by the finding that agricultural societies have, in general, been relatively incapable of generating economic growth on a scale comparable to that enjoyed by billions of people in the modern (post-1750) world. From the time of the "moral" Axial Age to the mid-eighteenth century, it is arguable that only one period stands out from the low energy-per-capita regime characteristic of agricultural societies that existed from the time of the invention of agriculture until very recently: namely, the heyday of the Roman Empire. If Ian Morris's calculations are at least broadly accurate, the Romans managed to increase the "energy capture" of their day significantly above what had been available before, thereby improving people's material well-being, nutritional status, and longevity. According to Morris, "Archeology makes it very clear that Romans ate more meat, built more cities, used more and bigger trading ships (and so on, and so on) than Europeans would do until the eighteenth century."[1] These centuries also witnessed the birth of Christianity, and Islam emerged about six centuries later, sweeping wide areas of the middle latitudes of the Old World in relatively short order.

In contrast to much contemporary understanding of Islam, however, Muslims "came not to bury the West but to perfect it."² After those innovations, however, one might argue that relatively little of a specifically religious nature emerged to rival the influence of the Axial Age religions and their Abrahamic heirs, Christianity and Islam. Max Weber's younger brother Alfred, a significant contributor to early twentieth-century German intellectual life in his own right, thus concluded that "since the [Protestant Reformation of the] 16th century, nothing fundamentally new has been added."³

Yet none of the developments associated with the first Axial Age had any dramatic effect on the overall amount of wealth that humans possessed. Most of the world's population continued to live at a relatively low subsistence level, with starvation and debilitating diseases a regular threat, while tiny elites enjoyed considerable advantages but still lived no longer than the rest of the population. By and large, little changed in this respect before about 1750. As the distinguished medieval historian Jacques Le Goff has recently noted, "Prior to the industrial revolution of the eighteenth century, there was only one real economy, based on agriculture."⁴ This world Fernand Braudel characterized as the "Biological Old Regime," "a set of restrictions, obstacles, structures, proportions and numerical relationships" that was "shattered in both China and Europe"⁵ only in the eighteenth century. The Biological Old Regime, according to Terry Burke, was an epoch when "periods of prosperity in which populations burgeoned, agricultural production soared, and trade flourished were followed by periodic subsistence crises in which famines and disease provoked downturns"; it "governed human affairs from the first millennium BCE. until ca. 1750."⁶ Even despite improvements during the later Middle Ages, the levels of productivity of pre-1750 Europe were "by our standards . . . still abysmally low."⁷ Similarly, the prominent economist Robert J. Gordon sums up the conventional wisdom among economists when he writes that "there was virtually no economic growth for millennia" before the late eighteenth century.[8]

At that point, however, economic development rose literally off the charts, creating much greater inequality than had existed previously[9] and spurring "the West" to take the lead over "the East." The Industrial Revolution then gathering steam would prove, to quote Gordon again, "by far the fastest and greatest transformation in history."[10] (See Figure 2.1.)

The period from approximately 1750 to 1850 was remarkable in so many ways that the German historical theorist Reinhard Koselleck dubbed it the "*Sattelzeit*"—literally the "saddle period," but more adequately translated as a "hinge" or indeed "axial" period. The latter translation seems especially appropriate insofar as Koselleck himself was mainly interested in the shift in historical consciousness or self-understanding that occurred during the era. In an early formulation, Koselleck wrote that he was interested in "the dissolution of the old world and the emergence of the new in terms of the historico-conceptual comprehension of this process." Above all, this new period is characterized by a rupture between experience and expectation and an awareness of the "contemporaneity of the noncontemporaneous."[11] In this sense, Koselleck's conception of the period shared the idealism of the thinkers who invented the first "Axial Age," who were focused more or less exclusively on its intellectual or ideological features.

Yet the term *Sattelzeit* began to acquire a number of other connotations over the course of time. In his magisterial history of the nineteenth century, Jürgen Osterhammel thus examines seven different dimensions of the *Sattelzeit* in the search for a globally relevant interpretation of the idea. These include: great power relations and the "first

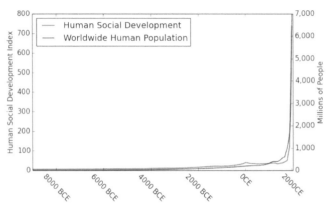

FIG 2.1. Human social development and population over time. Source: "Social Development," Ian Morris. http://ianmorris.org/docs/social-development.pdf; pre-1950 population stats from HYDE 3.1; post-1950 population stats from United Nations Department of Economic and Social Affairs.

age of global imperialism"; the beginnings of the political emancipation of European settler societies; the spread of nationalisms; the incipient but hardly effective inclusion of citizens in political decision-making; an emerging shift "from estate to class" and a rising challenge to slavery; the global takeoff of "modern"—that is, sustained and stable—economic growth on the basis of a new energy regime; and a sharply limited "globalization" in the realm of culture.[12] The push for women's equality, thoroughgoing political inclusion of subjects in "democratic" polities, the transformation of warfare into an activity principally carried out by the world's poor, full-scale industrialization of many places in the world—all this would await the twentieth century. Osterhammel's history of the nineteenth century is a valuable reminder that a great deal of what we think of as "modern life" is comprised of quite recent advances, often not more than a century old.

The developments that arose from the *Sattelzeit* as broadly construed have brought about enormous improvements in human well-being, but on the basis of a confrontational or neglectful attitude toward nature that has put the human species at considerable risk to human-induced climate change and global warming. It is as a result of these contradictory developments that the Nobel Prize-winning Dutch chemist Paul Crutzen and the biologist Eugene Stoermer argued that we should think of the current geological epoch as the "Anthropocene" era. They deployed this term to emphasize the unprecedented ways in which human beings have come to affect the environment and the resulting arrival of a new era on the geological time-scale. The idea of labeling the period since the late eighteenth century with a new name was intended to call attention to the challenge confronting the human race with regard to the continued viability of its ecological foundations. While the two authors understood that the choice of a starting date for these developments is intrinsically arbitrary, they originally suggested the latter half of the eighteenth century. They argued that "this is the period when data retrieved from glacial ice cores show the beginning of a growth in the atmospheric concentrations of several 'greenhouse gases,' in particular CO_2 [carbon dioxide] and CH_4 [methane]. Such a starting date also coincides with James Watt's invention of the steam engine in 1784."[13] From this perspective, the advent of the Anthropocene is essentially coterminous with the beginning of the Industrial Revolution, with its reliance above all on fossil fuels such as coal and, later, oil. More recently, however, there has been a push to date the Anthropocene as having begun after 1950 on the basis of data showing rapid increases in population growth and human-induced effects on the biosphere. This is the so-called Great Acceleration, a term that plays upon political economist Karl Polanyi's "holistic understanding of the nature of modern societies" as exemplified in such works as *The Great Transformation*.[14] Perhaps unavoidably, the matter of dating remains unresolved, but the post-1950 dating is gaining increasing currency.[15]

Notwithstanding its relationship to the dating of the Anthropocene, the Industrial Revolution had momentous and largely salutary consequences for the human species. Billions of people enjoyed more wealth, better health, and increased life expectancy. They were also relieved of backbreaking drudgery by steam and other sorts of nonhuman power, which could now do the work previously performed by many people. As a result, the cost of their basic necessities and the amount of working time necessary to purchase them both declined. Women's well-being and equality began to be taken seriously on a broad scale for the first time in history. The total wage-earning working time across the lifespan dropped dramatically, the proportion of working hours required on average to provide the basic necessities declined considerably, women's working hours fell, nutrition and health care improved, many previously fatal diseases were eradicated, life spans got longer (and longer and longer), and, especially since 1945, the proportion of deaths due to violent conflict fell.[16] For substantial segments of the world's population, life got dramatically better after 1750, although industrialization remained a relatively limited phenomenon, even in Western Europe and the United States, until at least the late nineteenth century.[17] (See Figure 2.2.)

In his *Why the West Rules—For Now*, Ian Morris displays a graph showing what he calls "social development"—"a group's ability to master its physical and intellectual environment to get things done"[18]—during the last two thousand years. The graph shows clearly that social development in both "East" and "West" was very low during most of this period, but it has exploded in the period since approximately

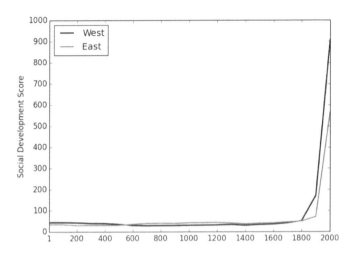

FIG 2.2. Social development by year. Source: "Social Development," Ian Morris. http://ianmorris.org/docs/social-development.

1750. In the process, "the West" distanced itself from "the East," becoming much wealthier by comparison. Why did this happen? The consensus among economic historians in recent years has emphasized a confluence of factors that promoted innovation. The Nobel Prize-winning economist Angus Deaton, who observes a similar improvement in the West during the post-1750 period, stresses that "new knowledge, new inventions, and new ways of doing things are the keys to progress."[19] Kenneth Pomeranz argued in *The Great Divergence* that this great leap forward had mainly to do with Britain's exploitation of the raw materials of North America and its own access to coal, permitting it to exploit the possibilities of James Watt's new steam engine.[20] Drawing on the work of Jared Diamond that emphasized the role of germs and the transfer of ideas among different ecosystems, Morris adds the advantages of geography to the explanation of "the rise of the West." What the East lacked was not merely, as Max Weber famously argued, a "Protestant ethic": "To have had its own industrial revolution, the East would have needed to create some equivalent to the Atlantic economy that could generate higher wages and new challenges, stimulating the whole package of scientific thought, mechanical tinkering, and cheap power."[21] Acemoğlu and Robinson stress above all the development of "inclusive" institutions rather than "extractive" ones.[22] Others would emphasize that the "Atlantic economy" involved the transfer of millions of Africans to the New World as slave labor. Osterhammel suggests that no general theory of industrialization is possible, but he offers a series of "key factors" that include those mentioned above and, perhaps in a nod to Weber, "a decidedly entrepreneurial spirit among small circles, especially of religious dissidents."[23]

What is most striking in this literature is the extent to which the "great divergence" has been pushed back from a putative sixteenth-century "rise of the West" (William McNeill) or of a Western-dominated "capitalist world-system" beginning in the sixteenth century (Immanuel Wallerstein) to a global economic takeoff after 1750 that led the West to outstrip the East—a position the West has maintained until fairly recently but whose relative weakening is now very much a matter for speculation. In short, the Industrial Revolution has returned to the center of discussion of modern economic development, but what stimulated it and when it led the West to get ahead of the East has been pushed forward to a much more recent date than had previously been emphasized. "Modern"—that is, more or less continuous—economic growth began with the Industrial Revolution in Britain and spread to Europe and the United States during the nineteenth century at the earliest; it did so because of the application of scientific technology to economic development.[24]

One of the most important testaments to the ability of modern societies to grow economically was that world population grew ... and grew, and grew, and grew. With advances in science and medicine, the planet came to support enormous numbers of people subsequent to 1750. In recent centuries, world population has increased at unprecedentedly fast rates in historical terms. In the year 1700, for example, world population totaled something under seven hundred million; it is now more than ten times that figure,[25] and it promises to continue to grow considerably in the decades to come. The parts of the global population that are large now will grow faster than those in the wealthier parts of the globe, if—with declining birthrates—at a slower rate than they have grown in recent decades. The center of gravity of the world's population will continue to be overwhelmingly in Asia, especially in China and India, which between them account for more than a third of the world's total population (some 2.6 billion out of 7-plus billion).[26] With greater wealth and improvements in health, life expectancy will continue to grow around the world, but differential fertility and mortality rates will mean that the rich world will gray faster while the poorer world will remain comparatively younger on average. Still, fertility rates will decline even in the poorer parts of the world as they become richer and their populations more educated, which in turn will gradually lead to older populations on average there as well.[27]

The population growth described in the foregoing has gone hand in hand with declines in fertility and age-specific mortality, which, when they occur together, are known as the "demographic transition." Since the beginning of the twentieth century, fertility rates have declined dramatically in the wealthy parts of the world, and they will decline in the poorer parts of the world as they (and especially their female populations) become wealthier and more educated.[28] This suggests a gradual slowing in world population growth, but from relatively high rates in some places and of course not nearly enough to lead to population decline overall. This process is essentially what is now taking place in China, where the population is aging and slowly declining in its *rate* of growth but not in absolute terms.

The big news with regard to longevity in human populations has been that, since roughly the beginning of the twentieth century, people in the wealthier parts of the world survive what had previously been deadly childhood illnesses (tuberculosis, polio, smallpox, cholera, and so on), live out their "natural" lives, and die of "old people's diseases"— especially heart disease, stroke, and cancer. This transition to living out one's full life has been much less in evidence in the world's poorer countries, however, and indeed this broad difference in mortality patterns constitutes one of the central axes of inequality in the world today. Still, the extension of life is affecting the poorer parts of the world as well, leading to dramatic gains in life expectancy worldwide.[29] While there were many centuries when advances of this kind did not reach the less well-off,[30] as Robert Fogel has put it, "if anything sets the twentieth century off from the past, it is [the] huge increase in the longevity of the lower classes."[31]

The economic improvements that have come to the world's poor are more recent, but have been extraordinary from the perspective of world history. While extensive poverty persists, of course, it has been reduced dramatically in recent decades—mainly as a result of a shift toward market-oriented policies in China and India. During the period especially since 1975, hundreds of millions of people escaped from extreme poverty as a result of these changes.[32] The Millennium Development Goal of achieving a 50-percent reduction in extreme poverty was reached in 2011, years ahead of schedule, leading to the adoption

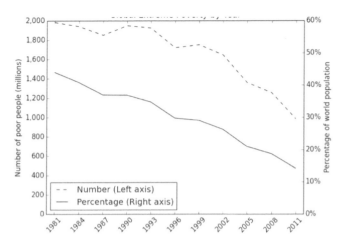

FIG 2.3. Global extreme poverty by year. Source: World Bank. http://research.worldbank.org/PovcalNet.

of a new panoply of Sustainable Development Goals. (See Figure 2.3.)

In short, industrialization—with all its costs in human discomfort and misery—has ultimately been very good for the world's population, especially its poorer parts. The paradoxical character of progress in recent centuries vindicates the tragic view of human history held by such writers as Walter Benjamin and Sigmund Freud; each believed that human progress was only possible on the basis of sacrifice. To be sure, the populations that made the Industrial Revolution had to endure the "dark Satanic mills" described by William Blake, but their hardship made possible a better life for billions of people who came after them.

One of the most important such improvements has been in the health of human populations. The chief contribution to the health of populations in the modern world has come from advances in public health (sanitation infrastructure, improvements in the water and milk supplies, vaccination campaigns, and so on). These advances came about mainly on the basis of government investment in public goods and infrastructure. For example, in the United Kingdom, major investments in sanitation in the late nineteenth century led to dramatic declines in infant mortality. In particular, the London sewer system, which came into existence some 150 years ago, made a huge difference in the health of those who benefited from it. Meanwhile, advances in water purification in the United States helped eliminate cholera and typhoid fever, major sources of waterborne disease until the late nineteenth century, and made a huge contribution to increasing life expectancy in the twentieth century. Indeed, according to David Cutler and Grant Miller, approximately half of the increase in life expectancy among city dwellers in the United States in the first half of the twentieth century as well as three-quarters of the reduction in infant mortality and two-thirds of the decline in child mortality resulted from treatment of drinking water.[33] Such investments in twentieth-century East Asia have significantly improved health and longevity there as well. In the absence of such investments, some 2.36 billion people around the world still live without basic sanitation and nearly one billion defecate in the open; the resulting disease kills more than half a million children every year—9 percent of all deaths among children younger than five years of age.[34] Assessing the contribution of these sorts of investments to longevity, Angus Deaton concludes that "the major credit for the decrease in child mortality and the resultant increase in life expectancy [during this period] must go to the control of disease through public health measures."[35] Needless to say, differences in exposure to life-threatening diseases resulting from lack of sanitation comprise a major disparity in the well-being of the world's rich and poor populations. Such infrastructure should be seen as a major element of social citizenship, one that is crucial to the integration of the lower classes into the mainstream of society.

In part because many infectious diseases can be controlled relatively easily and inexpensively, control of mortality from such diseases will surely improve in the poorer parts of the world in the years to come. For example, malaria, long a scourge of global populations, seems firmly set on the path to eradication.[36] Africa recently went an entire year without a reported case of polio, suggesting that the disease may be under control there. The problem had generally been in Nigeria, but then "officials embraced a vigorous new approach to vaccination and surveillance in that country," resulting in a year without reported infections.[37] Religiously motivated attacks on medical personnel have stymied vaccination campaigns in Pakistan, as a result of which that country faces persistent danger from the disease.[38] On the whole, however, dramatic progress is being made in regard to mortality from communicable diseases in the developing world.

In the face of these advances, deaths from infectious diseases such as malaria and tuberculosis have declined and the "global burden of disease" is shifting toward so-called diseases of affluence and of old age—in particular, diabetes and cancer. The rise of diabetes and cancer as causes of disability and death in the developing world is a product of aging and economic betterment, which may be seen as good things: "The growth of these rich-country diseases, like heart disease, stroke, cancer and diabetes, is in a strange way good news. . . . It shows that many parts of the globe have largely overcome infectious and communicable diseases as a pervasive threat, and that people on average are living longer."[39] Yet the increase in diabetes is the product of a rise in obesity, that poison pill of spreading prosperity. These trends both reflect enhanced life expectancy and thus contribute, along with declining fertility, to the relative aging of populations around the world—even ones that are currently young relative to the world as a whole. As a result of improvements in economic well-being and in life expectancy, the populations in the poorer parts of the world increasingly suffer the diseases typical of the developed world. This development is an example of the irony of affluence and the contradictory consequences of progress.

Still, assuming that recent trends continue, the world's people as a whole will get richer, while the distribution of wealth and income appears set to remain sharply unequal for a long time to come. Thomas Piketty has shown that inequality in the richer parts of the world is returning to levels previously seen before World War II, heightening our understanding of the anomalous character of the years the French call "*les trentes glorieuses*" ("the thirty glorious years") between 1945 and 1975, and that economists have dubbed "the great compression" in the United States.[40] From the perspective of the world as a whole, however, there appears to have been a growing convergence in global incomes—as measured across individuals rather than countries—in recent decades. According to Branko Milanovic, between 1988 and 2008 "perhaps for the first time since the Industrial Revolution [which distanced the industrializing West from the rest], there may have been a decline in global inequality."[41] Much of this has to do with the fast rates of growth that have occurred in China and India during this period. But it is a broader trend, reflected in the rise of the other BRICS countries (in addition to India and China, these are Brazil, Russia, and South Africa) and others, where economic growth has brought considerable improvement since the collapse of the Soviet Union.[42] Thomas Friedman's "flat world" thus has something to be said for it, as those countries that are getting richer also loom more significantly on the world political stage than they had previously done.

In addition to enhanced wealth, health, and longevity, the period since the mid-eighteenth century has also witnessed a major improvement in regard to the risk of violent death. War before that time was virtually all "guerrilla warfare"—hit and run, in effect, or, if there were fortifications, besiege and beleaguer. In the relatively infrequent cases when they were caused directly by battle rather than by diseases and untreatable wounds, combat deaths were brutal and face to face. Guns began to change that in the sixteenth century, but they were not terribly accurate or effective until the eighteenth century. The "age of democratic revolutions" then promoted and witnessed the "massification" of warfare, the raising by nation-states of huge armies of citizens prepared to die for the colors of the flag. The *levée en masse* during the French Revolution was the signature development of the period, declaring that all French citizens were to be part of the effort to defend the revolution from its detractors. Once warfare became an activity for conscripted citizens, mercenaries were no longer considered legitimate.[43] Increasingly in nineteenth-century European societies, soldiers fought in disciplined units rather than in rag-tag raiding bands. By the twentieth century, massed armies were joined by aerial bombing that could level whole cities. But the states' monopolization of the legitimate use of violence had come to be widely regarded as the norm, even if it was by no means achieved consistently within or across states.

As nation-states gradually took over from empires and monopolized the legitimate use of violence, rates of violent death declined dramatically in the post-1750 period. Rates of homicide in Europe have diminished substantially in the last several hundred years—perhaps one hundred-fold—with the steepest declines occurring since around 1700.[44] The extent and causes of these declines are varied according to time and place; the state's monopolization of the use of force has been less uniform and pervasive in the United States than it has been in Western Europe, and it has been less thoroughgoing in those parts of the United States that were beyond the reach of the federal government in the pre-Civil War period and for a time thereafter.[45] Still, the state's enforcement of its exclusive right to wield violence has been seen by prominent analysts as the chief restraint on violence and, along with the rise of a commercial society, the major contributor to declines in violence.[46] Since World War II, the use of force has declined other than in what one might think of as the "global ghetto," pacifying the rich parts of the world but making its poorer precincts more and more dangerous. Indeed, the militarization of European society before 1945 gave way to a more or less stable, nuclear-tipped peace that turned a previously bloody continent into a region about which one leading historian wondered, "where have all the soldiers gone?"[47] But even in other places, rates of violent death are generally significantly lower than they were in the pre-1750 period. The modern world may be the safest, most pacific world that human beings have ever known, particularly if one keeps in mind the enormous population growth that has taken place during that time. Yet serious dangers lurk in the shadows of the terrorist mind, if one or more of them get their hands on weapons of mass destruction.

These trends—demographic expansion and aging of the world's populations, greater life expectancy, growing wealth, and the decline of violence, especially in the world's wealthier precincts—are hallmarks of the era created by the Industrial Revolution. Notwithstanding the positive aspects of these trends, however, they have come at the cost of extensive environmental damage leading to warmer temperatures, volatile weather patterns, melting of the polar ice caps, acidification of the oceans, the extensive blanching and death of the world's coral reefs, rising sea levels, and more. The third

Axial Age has to reverse these trends if Earth is to remain habitable by the human species as well as millions of other species, which are dying off at extraordinary rates.[48] The second Axial Age vastly enhanced the health and material well-being of a larger and larger world population, but it must now learn how to continue to do so using more "sustainable" methods.

3
The Mental Axial Age

If the "material" Axial Age is defined by "more, more, more," the third, "mental" Axial Age is about using "smart" technologies to get more and more from less and less as well as developing technologies that produce thinking of their own, potentially cutting humans out of the loop. Automation, robotics, genetic engineering, and other such technologies have become increasingly pervasive, driving much social and economic change. It is a period that requires the introduction of a new conceptualization of and term for reality—"virtual" reality—because that technology "doubles" the old, traditional version, creating a new, fictional experience that may nonetheless seem as real as the "real" kind. Given newfound capacities for manipulating time and experience, meanwhile, we reinforce our old conception of time by invoking the appellation "real time." Even more, we have created machines with many of the capacities previously associated with and thought unique to human intellect; in addition to the old-fashioned kind, we now also have "artificial" intelligence—thinking that produces thinking, or at least purports or promises to do so. Finally, and perhaps most decisively from a sociological perspective, we speak of a divide between "digital natives" and their older predecessors.

Digital natives are those who have grown up with these new technologies, take them for granted, and experience mild bemusement and annoyance as they observe the perplexity with which their parents and other older people confront the "virtual" technologies that they themselves operate unproblematically. Daniel Bell foresaw much of this trend in social change when he wrote *The Coming of Post-Industrial Society* in the mid-1970s, but Bell did not emphasize the importance of climate change and global warming as strongly as now seems appropriate.[1] The third Axial Age is the age of "smart" things produced with less energy, and especially with other *forms* of energy, than those of the second. Arguably, they must do so in order to feed a growing global population and to save the planet from the unintended environmental consequences of the Anthropocene era.

The current third Axial Age is marked above all by various forms of "high technology." While there is no reason to insist on this kind of precision, the third Axial Age might reasonably be said to have begun in the mid-1970s, as the "Fordist" regime of assembly-line mass production ran headlong into the first oil embargo, a computer company called Microsoft was founded by a Harvard dropout tinkering in his parents' garage, and income inequality and wage stagnation among ordinary people began their inexorable rise in the United States—far and away the leading developer of these new technologies. Surely these phenomena are related to each other; one source of this wage inequality has been "skill-biased technological change," which has negatively affected the wages of those with less education and training as compared to the midcentury period of the "great compression."[2] Yet this has not been the only cause of wage stagnation; analysts also point to the effects of declining private-sector unionization, the falling real value of the minimum wage, offshoring of manufacturing production and the substitution of imports for domestic goods, as well as immigration.[3] In other words, "globalization"—which by measures other than international migration rates was quite high in the period following the collapse of the Soviet Union in the early 1990s—has led to declining economic prospects for the lower middle classes of the rich world, while improving the prospects of the "global middle class."[4]

Yet like the first Axial Age (at least if Baumard and colleagues are correct about its causes), the current epoch enjoys advantages of economic growth that make scarcity and starvation much less pressing problems for many people than they had been before the second Axial Age—a "postmaterialist" and, one might even very cautiously say, a postscarcity epoch, at least for a not-insignificant portion of the world's population. The period thus reveals contradictory tendencies that are crucial elements of its character—growing global technological development, wealth, and well-being in many respects, but also challenges to universalistic social provision in the wealthiest countries combined with a deterioration in the position of particular groups that had done well during the second Axial Age, and especially during the "great compression." It is these groups at the 80th percentile of the global income distribution, the working and middle classes of the wealthy countries of the world, who are experiencing economic and social dislocation and driving a politics of fear and nostalgia for the "better days" of the mid-twentieth century.

What is the shape of the economic changes that are taking place on the basis of the technologies of the third Axial Age? Some—let us call them the techno-optimists—argue that the quintessential feature of the third Axial Age is the exponential character of technological change and its revolutionization of everyday life. Just as population growth increased exponentially during the Great Acceleration, the techno-optimists argue, we are now experiencing an unprecedented intensification of change in the ways things work and in the way we live. The military affairs analyst P. W. Singer has described the shift in these terms:

> For the period up to the Industrial Age, the overall weight of technologic [*sic*] change was so slow that no one would significantly notice it within their lifetime. . . . By the late 1800s, change was playing out over decades and then years. . . . But this change period was just the start of an acceleration up [an] exponential curve. The current rates of doubling mean that we experienced more technologic change in the 1990s than in the entire ninety years beforehand. To think about it another way, technology in 2000 was roughly one thousand times more advanced, more complex, and more integral to our day-to-day lives than the technology of 1900 was to our great-grandparents.[5]

Against this background, *New York Times* tech columnist Farhad Manjoo has suggested as a New Year's resolution for 2016 that we "begin to appreciate the dominant role technology now plays in shaping the world."[6] The "mental" Axial Age is exemplified in physical terms by "Moore's Law": the axiom (not a true physical law of nature) that computing power doubles roughly every eighteen months at the same cost, leading to exponential growth in computational capacities and making possible a new era of "brilliant technologies."[7]

The techno-pessimists tend to object to the "whizz-bang" character of the optimists' attitudes and their tendency to think that we stand on the brink of a hitherto-undreamt golden era. They argue that the rapid doubling of computing power at the same price required by Moore's Law has been failing to put in its appearance in recent years. They stress the problematic consequences for employment associated with the recent development of high technologies, which compares unfavorably with the inventions of the late nineteenth and the first part of the twentieth century and their undergirding of a middle-class, consumer economy. They note that the technologies in question have mainly been confined to the realms of information, communications, and entertainment, and hence they have not had the broad economic impact of the large manufacturing industries we associate with the "Great Compression." Robert Gordon, in particular, argues that growing inequality and the flagging improvements in education and productivity with which it is associated comprise "headwinds" that will work against a repetition of the middle-class prosperity characteristic of the middle third of the twentieth century. From this perspective, the problem with the new technologies was summed up in a quip by the Nobel Prize-winning economist Robert Solow, who noted in 1987 that "we can see the computer age everywhere except in the productivity statistics."[8]

One might lodge three objections to the arguments of the techno-pessimists. First, contrary to their portrayal by the pessimists as blithely unconcerned about the downsides of contemporary technological development, the techno-optimists are aware that inequality is a major problem in contemporary American life. Although Brynjolfsson and McAfee are notably boosterish about the prospects of the "brilliant technologies" for enhancing human welfare, they highlight the growing socioeconomic "spread" to which those technologies are contributing and offer policy recommendations intended to counteract it. They are especially enthusiastic about a strengthening of the Earned Income Tax Credit—a sort of negative income tax that has wide support across the American political spectrum because it incentivizes work. Second, the pessimistic view of the potential contribution of new technologies to economic growth may not reflect the extent to which the new technologies improve well-being, because in many ways this cannot be measured in the GDP statistics. As *The Economist* noted recently, "Zero-priced goods are excluded from GDP . . . [as] are all voluntary forms of digital production."[9] This observation is a reminder of the many shortcomings of GDP as a measure of well-being; the efforts to develop more useful and compelling yardsticks for human welfare range from the Human Development Index constructed by Amartya Sen and others to Bhutan's famous measure of "Gross National Happiness." While GDP is clearly relevant to well-being, it falls short in many ways as a method of grasping how populations are actually doing, and the failure to measure the value of so much of the activity taking place online seems a significant distortion. Finally, both pessimists and optimists tend to have a gloomy view of the prospects for job growth in the face of ongoing technological advancement, although the techno-optimists tend to see something more like a "jobless future" resulting from automation.[10]

But the results on this front are likely to be mixed. The labor economist David Autor has shown that this is not the pattern revealed by the history of technological development. Instead, although some routine jobs will be eliminated, humans and technologies tend to interact; the result is that there are new jobs requiring judgments only certain kinds of skilled workers can make. Other studies support these arguments as well.[11] In the meantime, however, the tech giants of Silicon Valley have begun to take these issues seriously and have started meeting to map out plans to regulate the social consequences of artificial intelligence, with a particular eye on jobs.[12] Still, this is all likely to take some time to play out. Robert Allen has observed that computers are a "general-purpose technology" (GPTs), and that other such technologies such as the harnessing of steam power and electricity took many decades to pay off economically.[13] One might therefore say of the economic consequences of computerization what Chinese foreign minister Zhou Enlai is supposed to have responded to a question in the 1970s about the significance of the French Revolution: "It's too soon to tell."

Contemporary technologies increasingly fuse the human and the mechanical, the physical and the biological, the natural and the artificial. Of course, much of the Industrial Revolution was based on crossing boundaries in this way as well. But the Industrial Revolution produced things, not machines with "minds" of their own, and it cannibalized nature in pursuit of "more, more, more"; today, technology aims to work with nature, mimic it, and take advantage of its beneficial properties. Writing about a new gene manipulation technique called CRISPR, *New Yorker* journalist Michael Specter writes that the scientists he was examining realized that "if nature could turn these molecules into the genetic equivalent of a global positioning system, so could we."[14] A new, artificial means of introducing immunity into certain kinds of alien genetic material was thus born. Meanwhile, the scientists working on artificial intelligence are melding human beings with technologies in ways comparable to the ways in which biologists are implanting new capacities in genes. Already in 2001, a young man paralyzed from the neck down had a computer chip implanted in his head; as a result, he was able to send e-mail, draw, play video games, and change the channels of his television.[15] More recently, a quadriplegic with a chip in his brain was able to bypass his spinal cord and move his arms, though only if hooked up to a computer.[16] Similarly, the military wants to use artificial intelligence to create "Centaur Warfighting—systems that combine A.I. with the capabilities of humans, resulting in faster and more accurate responses than humans alone could achieve."[17] *The Six Million Dollar Man*, a futuristic 1970s television series starring a "bionic" man who fights evil and defends the good, has become the future of soldiering. The image of the cyborg seems increasingly indicative of the direction in which we are headed.

Human beings increasingly manipulate the tiniest particles of the natural world and turn them into a medicine, an energy source, or a better kind of food. And they do so with less and less energy, relying increasingly on renewable sources of fuel. As global population grows, they will have to do so if the increasing number of hungry mouths is to be fed in a sustainable fashion. Feeding nine billion people by the middle of the current century, the number predicted to be alive then, will require a 70 percent increase in food calorie production. This level of output is likely to require a major rethinking of the way in which food is produced. That is particularly true as the global middle class grows and begins to adopt the diets of the wealthier parts of the world. In the process, obesity, diabetes, and other diseases of affluence have become more widespread among the world's populations. Indeed, there is increasing recognition that the food system and the biosphere are strongly interdependent and that sustainable production of the amount of food necessary to feed the world's population will demand a shift away from resource-intensive meat diets and toward more fruits, vegetables, legumes, and nonfatty fish.[18] In order not to continue damaging the biosphere in the ways associated with the production of a meat-centered diet, this food will have to be grown in ways that lead to less pollution.

In general, people are being encouraged to produce with less and less waste; a "zero-waste economy" is one new mantra of the age. In the Anthropocene era, such waste has tended to be regarded as a mere "externality," to be disposed of without overmuch consideration of the harms that might be caused "downstream." This stance led to tremendous amounts of air and water pollution, but perhaps the chief result was the global warming that besets the planet today. The world's countries pledged in Paris in late 2015 to limit the global temperature rise to 2 degrees Celsius (3.6 degrees Fahrenheit) above its level before the Industrial Revolution, but global temperatures have already risen by about half that amount.[19] In addition, the planet teeters on the brink of carbon dioxide concentrations of 400 ppm, a level above which reducing the level of CO_2 in the atmosphere may become impossible.[20] Techno-optimists such as economic historian Joel Mokyr believe that a technological fix to the problems of climate change is possible, "even if nobody can predict right now what that will look like, or if collective action difficulties will actually make it realistic."[21] Former German Green Party cochairman Ralf Fücks makes much the same point, arguing that while there is no guarantee that new technologies will actually stave off the consequences of global warming, they are essential tools if we are to head off the potentially catastrophic effects of the carbon-based energy regime that has fueled the Anthropocene era.[22]

In addition to producing more and more with less and less, the new information and communications technologies of the third Axial Age are arguably destroying less by making warfare an increasingly "high-tech" affair. Yet there have been a number of causes leading to the decline in deaths due to armed conflict in the post–World War II era. The atomic bomb was the culmination of the era of mass warfare, with the introduction of an awesome technology that made possible instantaneous mass death. Its unprecedented capacity for destruction seemed to many to make war among major powers unthinkable. Nuclear weapons thus contributed to the advent of an extraordinarily peaceful Europe after 1945, with occasional and—compared with the period before 1945—relatively minor exceptions in the Balkans and

Ukraine. With militaries in the developed world more and more based on all-volunteer forces (with the burden of military service falling mainly on the less fortunate), the era of large-scale warfare involving major proportions of a country's population seems to have come and gone in the richer parts of the world.[23]

One of the most striking developments during this period has been the decline in the number of interstate conflicts—war in the old-fashioned sense, defined as an event where two or more states fight each other—almost to the vanishing point in recent years. In a 2015 study of patterns of conflict in the post-World War II period, two leading peace researchers counted only one interstate conflict in 2014, that between India and Pakistan, "which led to fewer than 50 fatalities."[24] As the nation-state form came to be the accepted norm in state organization, the conquest of other states simply came to seem illegitimate and has largely disappeared from the map. The era of decolonization bore witness to this global shift toward the norm of nation-state organization. Hence, as Andreas Wimmer has put it, "The transformation of the nature of states . . . helps [us] to understand why wars between states have become so rare in the contemporary period."[25]

Instead of old-fashioned interstate wars, the era has been widely seen as one of "new wars" that resemble those of the medieval period, with warlords squaring off against one another over resources, rare minerals and gems, and women; they often seem to be forms of, or at least deeply implicated in, organized criminal activity as much as groups with clear political aspirations. Now most "hot" warfare takes place within states in the poorer parts of the world, mainly pitting rival ethnic groups against each other with machine-guns mounted on the back of pickup trucks. Little artillery, few tanks, hardly any aerial bombing (Syria in the conflict's later stages being the exception that proves the rule); hence the body counts are rather low by historical standards.[26] Since the Second World War, the number of people who die in battle in all types of armed conflict has declined dramatically, especially since the proxy wars between the capitalist and communist worlds fell by the wayside with the demise of the Soviet Union in the early 1990s. This trend appears clearly in the following graph depicted in Figure 3.1.

After 1945, computers became the principal nanotechnology contributor to warfare, but their significant impact on the conduct of war has been relatively recent. According to P. W. Singer, the Gulf War of the early 1990s was the first to involve computers to a substantial extent. Computers helped organize troop movements, identify aerial bombing targets, and guide bombs to their destinations.[27] In the meantime, more and more robots are deployed to do dirty, dumb, and dangerous work such as mine clearing, identifying ambushes, and even taking out certain targets. Soldiers love them because they save their lives. Drones—unmanned aerial vehicles or UAVs—are "piloted" sometimes from thousands of miles away and are increasingly used for surveillance or for bombing enemies. They will of course also soon become a much more significant fact of civilian life. If Jeff Bezos has his way, drones will also soon be delivering packages for Amazon. Fire departments will use them to get more information about their enemies—fires. Finally, the FAA recently required that recreational UAV's of one-half pound to fifty-five pounds—of which it estimated that some seven hundred thousand would be sold around Christmastime 2015—be registered so it could keep track of their ownership.[28] New regulations will surely be required to insure that these new technologies, whose main use today is military, will be used safely on the domestic front.

In keeping with the shift toward high-tech military operations, military expenditures by the world's wealthier

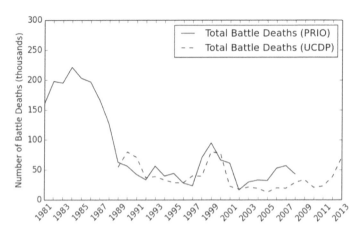

FIG 3.1. Battle deaths over time. Source: Peace Research Institute Oslo (PRIO) "(PRIO) Battle Deaths Dataset Version 3.0 (October 2009) and Uppsala Conflict Data Program "UCDP Battle-Related Deaths Dataset v.5-2014, 1989-2013," (June 12, 2014), Uppsala University, Uppsala, Sweden.

countries are shifting away from labor-intensive armies in favor of huge investments in cyber warfare and tiny, elite special operations forces. In the annual "threat assessment" presented by American intelligence officials to the Senate Intelligence Committee in early 2013, the director of national intelligence argued that the risk of a major cyber attack was a greater immediate threat to the country than international terrorism—the first time any other threat had been ranked more highly than terrorism since 9/11.[29] A new cyber "cold war" has emerged between the United States and China in recent years as the Chinese have developed their cyber weapons as well.[30] Despite a broader climate of defense budget cutting, cyber warfare is one of the few areas where spending is likely to grow in the years to come.[31] As warfare becomes more technology-based and precise, new ethical questions about the use of force arise and conflicts between the developers of new technologies and their military users may grow.[32] But these new technologies arguably reduce the numbers of those dying in battle and—assuming they do not make it easier for their wielders to use force inappropriately[33]—enhance human well-being by reducing the numbers of people dying in armed conflict. In sum, even warfare has gone high-tech because the populations of countries in the rich world tend to be more averse to combat deaths than they once were, and the newer technologies promise both precision killing of the enemy and minimal casualties on one's own side.

In part as a result of declines in violence, life expectancy has generally risen around the world over the last century. But should we expect this pattern to continue in the third Axial Age? Some believe that, given the extraordinary increases in life expectancy in recent decades, we must be pushing up against biological limits to life expectancy. In contrast, Oeppen and Vaupel have argued that global peak expectation of life increased on average three months per year over the previous 160 years, or a total of forty years' increase since the mid-nineteenth century.[34] Although Nobel Prize-winning economist Robert Fogel admits that he leans toward the optimistic, he and his colleagues incline to the belief that the more conservative forecasters of life expectancy have consistently underestimated the potential for further longevity gains, and that average life expectancy in the developed world, at least, will grow for some time to come.[35] Even countries that were poor not many years ago have shown stark improvements; in Brazil, for example, life expectancy has grown from 62.5 years in 1980 to 74.9 years in 2013, a gain of almost 20 percent in a generation.[36] Still, the post-Soviet collapse of men's life expectancy in Russia is a powerful reminder that, under certain conditions, the trend can go in the opposite direction rather quickly. Facebook founder

Mark Zuckerberg and his wife have recently announced that they plan to devote tremendous resources to eradicating the chronic diseases such as heart disease and cancer that afflict the increasingly aging population of the world; after all, Zuckerberg averred, "Medicine has only been a real science for less than 100 years, and we've already seen complete cures for some diseases and good progress for others. . . . As technology accelerates, we have a real shot at preventing, curing or managing all or most of the rest in the next 100 years."[37] The extraordinary optimism of the tech elite could hardly be more evident than in this statement. Some are wondering: Can we live forever?[38]

Yet extended expectation of life does not necessarily imply greater *healthy* life expectancy. Even if its rise has slowed recently, the average age at which chronic diseases have set in rose steadily during the course of the twentieth century, foretelling continued increases in healthy life expectancy. However, recent research has shown worldwide rates of years lived with disability declining more slowly than mortality rates, which points to the need for more attention to the chronic, nonfatal diseases that people tend to live with in old age.[39] In general, as global populations age, accommodations will have to be made in order to achieve what the World Health Organization (WHO) has termed "functional ability." This concept includes two dimensions, one concerning the "intrinsic capacity" of the aging person, the other referring to the environment in which such persons must live their daily lives. One major implication of increased aging in populations is that the traditional focus on curing disease tends to be inappropriate for older populations because it "detracts from helping older adults to get the help they need to improve their functioning irrespective of the multitude of health disorders they might have."[40] The shift from maladies requiring acute care to those demanding long-term care for older people is paralleled by the shift in the burden of disease from communicable to chronic diseases even in the non-wealthy parts of the world. This transition requires a reorientation of public health programs in these countries, which have mainly been focused heretofore on eradicating communicable and infectious diseases.[41] Another implication of these trends is that the people who, in older age, require the most assistance tend to be the people with the fewest resources on which to call to cope with their difficulties.[42] More arrangements for long-term care will have to be created and more labor power devoted to caring for the infirm elderly, yet much of this will have to be paid for by governments. This set of issues is likely to demand a major infusion of resources, including in retrofitted infrastructure to facilitate movement by less-mobile people, in coming decades. Whether such spending can be expected is not clear; a recent study suggests that, while global spending on health is likely to more than double between now and 2040, expenditures in the world's poorer countries are unlikely to grow adequately to meet the need that is likely to arise.[43] The following graphic summarizes much of the above information by showing the patterns and causes of deaths worldwide in recent years, as well as the ages at which people die in different income groups around the globe. (See Figure 3.2.)

It should also be noted that the aging of the world's population is not purely a matter of having many more infirm people who will drive up health-care and other costs. For example, there is good reason to expect that global population aging will yield a "gray peace dividend." A widespread consensus exists that so-called youth bulges—disproportionately large shares of young people aged fifteen to twenty-four in

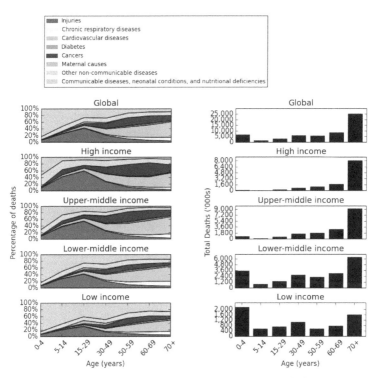

FIG 3.2. Deaths by World Bank income category: causes and totals. Source: World Health Organization Global Health Estimates 2014 (data for 2012 shown).

the adult population, especially where they lack employment opportunities—are the cause of much violent conflict.[44] It therefore stands to reason that, as the world's population ages, and especially if it continues to grow wealthier, such conflict will lessen. Against the background of insights such as this, the authors of a recent WHO study on aging and health argue that a fundamental shift is required in the way that we think about older populations. Far from being exclusively a burden, these people make all kinds of contributions to families, communities, and society. The fact that they are often no longer in the labor market should not distract us from the value of what they contribute. The *World Report on*

Ageing and Health stresses that older people need to be able to live with dignity, a prospect that will require changes in our attitudes towards the elderly, in the infrastructures that surround them, and in the kinds of care that they will need in order to thrive in old age.[45]

Much of the success in reaching these objectives will depend on investments in public health of the kind that yielded major improvements in life expectancy and health in the first part of the twentieth century. The incorporation of the working class into the health and longevity regime enjoyed by the wealthier segments of society rested heavily on improvements that advantaged everyone, not just the wealthy. This infrastructure includes the physical structures that make possible movement, consumption, and many aspects of people's daily routines; they are the sort of thing that T. H. Marshall had in mind when he wrote about social citizenship as entailing "the right . . . to live the life of a civilized being according to the standards prevailing in the society."[46] Yet recent evidence suggests that the hugely important historical trend toward public health advances that led to longer lives across class boundaries in the twentieth century may be going into reverse, at least in the United States. For example, the parlous condition of the physical and transportation infrastructure in the United States, as assessed by the American Society of Civil Engineers (ASCE) in its quadrennial report, thus constitutes a major threat to social citizenship and an important sign of the fraying of the social fabric. Replacement of the pipes that conduct drinking water into millions of American homes, for example, could cost upwards of a trillion dollars.

On the bright side, however, the ASCE finds that—cases such as Flint, Michigan notwithstanding—the drinking water system in the United States only very rarely leads to illness: "Even though pipes and mains are frequently more than 100 years old and in need of replacement, outbreaks of disease attributable to drinking water are rare."[47] Moreover, the country's waters have improved considerably since the Clean Water Act of 1972,[48] as has its air quality. Indeed, the quality of the air in the United States has improved faster than that of our Western European counterparts, which rely more heavily on dirtier diesel fuel. That said, people in cities in wealthier countries are generally much better off than those in poorer countries, where almost all urban settings with more than one hundred thousand inhabitants have levels of air pollution that exceed WHO standards.[49] Many commentators have pointed out that the post-recession climate of extremely low interest rates offers an ideal opportunity for investments aimed at modernizing public infrastructure, but political obstacles have tended to make this impossible in the United States.

The negative consequences of the failure to strengthen infrastructure that benefits the broad majority of the American population, combined with growing economic inequality, may be becoming clearer with new research findings. For example, Komlos and Baur found that the U.S. population had shifted "from the tallest to (one of) the fattest" in the past century and a half. Having been taller than Europeans in the mid-nineteenth century, Americans are now some three to seven centimeters shorter than their European counterparts; the Dutch, Swedes, and Norwegians are the tallest, while Americans have become the most obese population in the wealthy world (although Mexicans have higher rates of obesity). The authors suggest that this striking finding may be "related to the greater social inequality, an inferior health care system, and fewer social safety nets in the United States than in Western and Northern Europe, in spite of higher per capita income."[50] Similarly, the National Academy of Sciences recently found that, despite its wealth and its much larger relative outlays for health care, the United States lags behind other rich nations in health outcomes generally. Moreover, whereas the American population at midcentury was generally healthier than the populations of the countries with which it is normally compared, the health of the U.S. population is now falling behind those countries in many categories as well. The report also notes that the American health disadvantage relative to other wealthy countries cannot be attributed simply to disparities in health outcomes between racial or socioeconomic groups, "because recent studies suggest that even highly advantaged Americans may be in worse health than their counterparts in other countries."[51]

Focusing specifically on rates of death among different groups within the United States, S. Jay Olshansky and colleagues recently found that the disparities in life expectancy between well-educated whites and less-educated blacks in the United States had grown between 1990 and 2008, such that "at least two Americas have formed, with notably different longevity prospects."[52] Meanwhile, Raj Chetty and his colleagues found that, between 2001 and 2014, life expectancy in the United States increased by 2.34 years for men and 2.91 years for women in the top 5 percent of the income distribution, but by only 0.32 years for men and 0.04 years for women in the bottom 5 percent. A correlation was found between income and life expectancy across the range of incomes. More specifically, men in the bottom 1 percent of the income distribution at the age of forty had an expected age of death of 72.7 years, whereas men in the top 1 percent of the income distribution had an expected age of death of 87.3 years—a difference of 14.6 years or approximately 20 percent more than the life expectancy of those in the bottom 1 percent. Meanwhile, women in the bottom 1 percent of the income distribution at the age of forty had an expected age of death of 78.8 years, whereas women in the top 1 percent had an expected age of death of 88.9 years, 10.1 years longer than those in the bottom 1 percent.[53]

For their part, in research that generated a good deal of media attention when it appeared, the husband-and-wife team of Anne Case and Angus Deaton found that mortality rates of white middle-aged men and women in the United States have risen in the years since 1998, especially among the less educated. Remarkably, Case and Deaton found that these middle-aged whites were the only demographic group measured whose mortality rose during this period. (Their findings resonate with those of Olshansky and collegaues, who also found that less-educated whites were facing declining longevity prospects during the period they studied, 1990–2008.) They suggest that these findings may reflect the long downturn in the economic prospects of these populations and rising concern among them about economic insecurity since the 1970s. Disturbingly, Case and Deaton muse that "those currently in midlife may be a 'lost generation' whose future is less bright than those who preceded them."[54] In addition to these distressing findings, the rate of suicides in the United States increased by one-quarter between 1999 and 2014, with the rate of increase growing after 2006—which is to say, after the financial crisis that ensued shortly thereafter.[55] All these findings seem to point to the major public health consequences of the deterioration in economic security and social provision experienced by particular population groups in recent years. These disadvantages are especially pronounced among the less educated, which raises concern about the declining rates of educational attainment in the United States in recent years.

There are silver linings in this research, however. Olshansky and colleagues found that the proportion of blacks gaining higher levels of education rose during the period they studied (1990–2008), which promises to counteract the trend toward divergences in life expectancy between well-educated whites and less-educated blacks. Yet at a time when states are steadily decreasing their contributions to public higher education, this trend may be threatened, as minorities tend to rely more heavily on financial aid and on public institutions in order to attend college. Education is to some extent a reliable proxy for income and wealth, but may have independent effects as well. The general finding that those with greater education live longer and healthier lives suggests that more educational opportunities should be made available to people across the life course. More broadly, this research suggests, across-the-board improvements in health and education are more likely to promote healthy longevity than individual interventions.[56] Another countervailing trend is suggested by the findings of J. Currie and H. Schwandt, who found—in contrast to the worrisome trends among American adults—"strong improvements in mortality" among those under the age of twenty. Moreover, these improvements were most notable in the poorest counties in the United States, implying "a strong decrease in mortality inequality" between rich and poor.[57]

The trends in the United States are in some respects out of step with developments on the world scene, however. As a result of the gains in health and longevity worldwide, the world's population is projected to increase from its current level of approximately seven billion to perhaps nine billion or more by the mid-twenty-first century. The remarkable growth of population early in the period following 1750 led to concerns about the ability of food producers to provide enough to eat for the planet's inhabitants. Thomas Malthus famously articulated this fear of too many mouths to feed in his *Essay on the Principle of Population* (first published in 1798), which to this day remains a touchstone of debate about population dynamics and their consequences. Indeed,

Paul Ehrlich's neo-Malthusian 1968 book *The Population Bomb* foretold mass starvation as a result of the "population explosion." Economist Julian Simon, pointing out that the prices of most commodities had declined over time, wagered that this would also be true for any basket of commodities Ehrlich chose over a ten-year period. Simon won the bet.[58] Advances in food production have so far obviated these concerns, although there are certainly instances when the ability of particular populations to afford food has been outpaced by prices. So far, the Malthusian trap has generally been avoidable so long as politics in the relevant place are democratic. In his research on famines, Amartya Sen found that only undemocratic states—and no democratic states—have suffered famines.[59] Still, the prospect of feeding nine billion people while retooling the sources of energy used to produce food seems daunting indeed. New technologies, and perhaps a shift in global food preferences away from beef and pork, may well be necessary to manage this transition.

Despite the many positive aspects of recent technological development, by no means everything about these technologies is benign. While the risk of violent death may have continued to diminish during the third Axial Age, the potential for huge devastation remains and in some ways has gone underground. Any knowledgeable hacker may be in a position to wreak enormous damage on populations and infrastructures. As Brynjolfsson and McAfee put it in their treatise on today's "brilliant technologies," "We can reap unprecedented bounty and freedom, or greater disaster than humanity has ever seen before."[60]

On a less apocalyptic level, a longtime observer of the social consequences of new technologies, Sherry Turkle, has found that the potentially connecting characteristics of the new communications technologies have actually led to people being "alone together"—sharing a physical space, but only tenuously related socially because they are more engaged with what's going on in cyberspace than with what's going on in the room they are actually in. Young people—those "digital natives"—tend to text rather than to talk on the phone or meet up in person; such encounters entail emotional risks that many of them say they cannot face. Thus while the new communications technologies may make possible much greater connection to others, they may undermine actual conversation.[61] The spread of mindfulness meditation and other practices associated with contemplative Asian religions tracing back to the first Axial Age may be efforts to compensate for the anxieties arising from the new technologies of communication as well as the intensified pace of life accompanied by ubiquitous electronic devices.

There is another way in which the new communications technologies divide rather than unite. Digital natives find their elders' lack of facility with the new technologies amusing, but also alienating. "Why can't you do this?" my twelve-year-old daughter asks exasperatedly when I fail to use my iPhone with the efficiency with which she can use it (even though it's "mine," not hers). To be sure, this divide has a biological solution; before too long, everyone will be a "digital native," at least in terms of when they arrived on the planet (if not necessarily in the sense of having equal access to these technologies).

All this is likely, over the short- to medium-term, to generate conflict between the young and the old over resources. Yet the old tend to control political systems and are more likely to vote. The unaccommodated young may be forced to languish in limbo for a longer time than they had expected, wasting some of their potentially most productive years. The growth of employment in health care occupations offers some promise; many young people will need to be channeled into jobs providing for the needs of an older and older population—and the old will have to pay for these services. The organization HelpAge International is perhaps the leading organization trying to track and advocate for services that make for more congenial aging.[62] The growing numbers of elderly may themselves benefit from advances in technologies such as robots and drones—but at a growing cost to societies increasingly weighted in demographic terms toward the aged.[63]

Economist Robert Fogel has observed that, with greater longevity and declining work time necessary to acquire life's necessities, there may be many more opportunities for people to undertake voluntary work, as opposed to the kind that leads people to say, "I owe, I owe, so off to work I go."[64] Fogel presents this situation as an opportunity for self-fulfillment, but that depends on whether people have the wherewithal to engage in such activities. If everyone is living longer but not necessarily working for pay, they need income in order to continue to do the things they want to do and are capable of doing. But during the working years, the concern is growing whether *any* work will be available from which to earn a decent living. The vagaries of the relationship between work and income in the third Axial Age has thus given fresh momentum to the discussion about a "universal basic income," a long-standing objective of some on the left (and even on the right).[65] One difference now is that the idea is being explored seriously by the tech moguls of Silicon Valley. Following a largely ignored experiment in Manitoba during the mid-1970s, the tech incubator firm Y-Combinator is running its own experiment in Oakland, California, to determine whether the dire consequences predicted by critics—such as a loss of any work ethic, decline in meaningful activity, and budget-busting expenses—will actually ensue from a guaranteed minimum income.[66]

An additional concern is the "Frankenstein" fear associated with the laboratory creation of new "forms of life."

Artificial intelligence aims to design machines capable of intelligent reasoning. Many will know about the case of a computer that beat a previously very successful human player at *Jeopardy*, as well as about computers beating chess and Go masters at their own game. But artificial intelligence involves the creation of entities that some worry may escape their creators' intentions with regard to self-initiated decision-making and action. Those working on these issues are debating whether artificial intelligence can generate "existential threats" to the human species. What had once seemed a remote possibility largely scoffed at by serious scientists is now a matter of sustained reflection by a number of think-tanks around the world—the Centre for the Study of Existential Risk at Cambridge University, the Future of Life Institute connected to MIT, and the Future of Humanity Institute at Oxford University. The director of the latter, the philosopher Nick Bostrom, asks, "Will an A.I., if realized, use its vast capability in a way that is beyond human control?" Bostrom does not believe that the answer to that question is known, but he does insist that "artificial intelligence already outperforms human intelligence in many domains."[67] Mary Shelley's concerns of two hundred years ago now confront vastly more sophisticated technologies than she could have imagined. However, in an echo of the Judeo-Christian understanding of the apocalypse, some of the leading figures in the artificial-intelligence community express hope for eternal life after uploading their brains into computers and transcending the shortcomings of earthly bodies.[68]

If the "carrier groups" that bore the principal weight of the second Axial Age were the bourgeoisie and the proletariat, today they are the techno-libertarians and their financial-industry facilitators in Silicon Valley, Shanghai, Stuttgart, and Bangalore, on the one hand, and those who operate in what we might call the immobile economy of things and services, on the other. The techno-libertarians provide innovations that may lighten the burdens of those in the immobile economy or that may deprive them of jobs. Robots that sweep floors in homes may make life easier for those who have them, but robots that sweep floors in industrial factories reduce the manufacturers' labor costs by putting people out of work. Although socially liberal, the techno-libertarians and their financial counterparts tend to see government as an obstacle to be overcome and prefer to see entrepreneurs run everything, including the institutions—such as public schools—that provided the intellectual foundation of mass citizenship in the period since the mid-nineteenth century. They have also had a major effect on philanthropy, insisting as they tend to do in all realms of life that their charitable giving be data-driven, outcome-oriented, and accountability-based. Gradually, this approach is even having an effect on development assistance provided by governments as well.[69]

Those in the immobile economy who rely on those institutions may object, but they sometimes agree that things are run better by entrepreneurs than what may appear to be staid, underperforming institutions from a bygone era, such as public schools and teachers' unions. Citizenship is increasingly hollowed out as an institution that can be bought for a price by those who can afford it; those too immobile to purchase other, more congenial citizenships face a steady shift toward the privatization of citizenship-related services—of roads, police, water supplies, and the like, but perhaps above all of education, which is being offered more and more by private or for-profit providers from the United States to Germany to Turkey and beyond. For some in the poorer parts of the world, private education is nothing new and much preferable to the public alternative. But in the wealthier parts of the world this trend is a rollback of the mass investments of the second Axial Age that, like those in public health and sanitation infrastructure, enhanced the well-being of ordinary people in unprecedented ways.

It is thus well to remind ourselves that the new communications and information technologies of the third Axial Age may help pull the planet back from the brink of destruction, but they won't necessarily lead to the end of class disparities, racism, gender inequality, and the like. Indeed, there is a notable disjunction between the optimistic cast of thought among those pursuing solutions to technical challenges—whether these involve writing an app for practical everyday problems or the creation of machines that can think more effectively than humans—and those who concern themselves with social problems. Black Lives Matter and the Zero Campaign, two newly emerging social movements that focus on ending police violence against black people in the United States, remind us that we remain mired in social problems that seem immune from improvement by purely technological means.

The big question facing the human species at present is whether we can make the transition to a mode of production that restores a balance between the human and natural worlds while sustaining unprecedented numbers of people. There are encouraging signs that economic growth, which long marched in tandem with increases in atmosphere-warming carbon emissions, is possible without such emissions increases.[70] Inequality has grown noticeably worse within countries since the onset of the third Axial Age, but recent changes in the world economy have also pulled billions of people out of poverty and leveled the playing field *among* if not within countries. Still, according to the World Bank, the gains for the poor in recent years are threatened by global warming, which is likely to hit the world's poor hardest unless development strategies address climate change.[71] The trend toward greater inequality will not change without significant shifts in social policies designed to support a labor regime that will include greater "precarity" than that associated with the Fordist era of mass production. Still, the economic problems we face are to a considerable extent simply distributional, not a matter of trying to produce the necessary wherewithal. Negative income taxes are likely to be one of the more politically palatable ways to achieve the necessary redistribution to put the opportunities of the third Axial Age within reach of all.

Awareness of environmental degradation at the dawn of the third Axial Age led to regulations in the developed world that began to turn around the desecration of the planet. If contemporary technologies and the flattened organizational hierarchies they generate can make social equality a reality while pulling us back from ecological disaster, the era will have used the material resources provided by the second Axial Age and the moral values created by the first to usher in a period of exceptional human freedom and possibility. But we still have a long way to go before these goals are achieved. I am inclined to agree with *New York Times* technology writer John Markoff when he writes, after observing a self-driving car being developed in the Israeli suburbs, "This was the new Promised Land. . . . We are no longer in a biblical world, and the future is not about geographical territory, but rather about a rapidly approaching technological wonder world."[72] Whether it will be a sociological wonder world remains to be seen.

In all events, we will have to draw on the resources supplied by the first Axial Age in order to address these problems. But this need not be a cause for despair. Take, for example, Pope Francis's recent encyclical *Laudato Si': On Care for Our Common Home*. Francis takes central aim at the "technocratic paradigm" that he regards as the root of our inability to confront seriously the ecological challenge we face in global warming, and he argues against those who see in Christianity a belief system in which the Earth is simply handed over to humans for domination. Instead, he insists, "the Bible has no place for a tyrannical anthropocentrism unconcerned for other creatures."[73] The outspoken environmental activist and writer Bill McKibben has described the document as "a sweeping, radical, and highly persuasive critique of how we inhabit this planet—an ecological critique, yes, but also a moral, social, economic, and spiritual commentary."[74] Even if the cultures of the descendants of Axial Age civilizations have been chiefly responsible for creating a profound ecological morass, they may not be entirely incapable of generating an adequate response to that morass from within their own cultural-religious resources. Given that Pope Francis presides over what one might regard as the world's largest charitable organization—one that rejects the dominant political and economic tendencies of the day in favor of non-exploitive, nurturing relationships among humans themselves and between humans and the planet—this is a matter of no small import. For example, the head of a major electricity-producing company, Mauricio Gutierrez of NRG, is trying to generate more power from renewable sources such as solar, and he sees the Pope's message as important motivation for his efforts. "When a spiritual leader like the pope calls out our moral responsibility toward the environment, it's a pretty big thing," Mr. Gutierrez said. "It transcends science and policy."[75] The Pope has even proposed adding care for the environment to the seven works of mercy traditionally incumbent upon Christians, a significant revision of a doctrine with roots in the first Axial Age that is oriented to making that doctrine relevant to the present.[76] These resources will be essential to addressing the social and ecological challenges we face during the third Axial Age.

Notes

Preface

1. Robert Bellah, "Religious Evolution," in *Beyond Belief: Essays on Religion in a Post-Traditionalist World* (Berkeley: University of California Press, 1991), 20; this was previously published in *American Sociological Review* 29, no. 3 (June 1964), without the explanatory note from which I draw the quoted passage.
2. J. B. Bury, *The Idea of Progress: An Inquiry into its Growth and Origin* (New York: Macmillan, 1932).

Introduction

1. Among other contributions, see Robert Bellah, *Religion in Human Evolution: From the Paleolithic to the Axial Age* (Cambridge, MA: Belknap/Harvard University Press, 2011); and Robert Bellah and Hans Joas, eds., *The Axial Age and its Consequences* (Cambridge, MA: Belknap/Harvard University Press, 2012).
2. Jaspers developed the notion of the Axial Age in *The Origin and Goal of History*, trans. M. Bullock (London: Routledge & Paul, 1953 [1949]).
3. See John Boy and John Torpey, "Inventing the Axial Age: On the Origins and Uses of a Historical Concept," *Theory & Society* 42, no. 3 (May 2013): 241–59.
4. For a critical discussion of the notion of "world religions," see Tomoko Masuzawa, *The Invention of World Religions* (Chicago: University of Chicago Press, 2005).

5. Eric Brynjolfsson and Andrew McAfee, *The Second Machine Age: Work, Progress, and Prosperity in a Time of Brilliant Technologies* (New York: W. W. Norton, 2014).
6. Karl Marx, "The German Ideology," in *The Marx-Engels Reader*, ed. Robert Tucker, 2d ed. (New York: W. W. Norton, 1972), 160.

1. The Moral Axial Age

1. Ian Morris, *Why the West Rules—For Now* (New York: FSG, 2010), 88ff.
2. Marshall Sahlins, *Stone Age Economics* (Hawthorne, NY: Aldine de Gruyter, 1974).
3. For the negative health effects on humans of civilization, see Mark Nathan Cohen, *Health and the Rise of Civilization* (New Haven, CT: Yale University Press, 1989).
4. Ibn Khaldûn, *The Muqaddimah: An Introduction to History*, trans. Franz Rosenthal (Princeton, NJ: Princeton University Press/ Bollingen Series, 1967).
5. Ian Morris, *War! What is it Good For?: Conflict and the Progress of Civilization from Primates to Robots* (New York: FSG, 2014), 58–59.
6. S. N. Eisenstadt, "The Axial Age: The Emergence of Transcendental Visions and the Rise of Clerics," in *Comparative Civilizations and Multiple Modernities* (Leiden and Boston: Brill, 2003), 199. Eisenstadt was enormously influential in the discussion of the Axial Age, but—as this passage suggests—his writing can be maddeningly vague, abstract, and repetitive. For a critique, see Ian Lustick, "The Voice of a Sociologist, the Task of a Historian, the Limits of a Paradigm," in *Books on Israel* (1988), 1: 9–16.
7. Max Weber, *Economy and Society*, ed. Guenther Roth and Claus Wittich (Berkeley: University of California Press, 1978), 1: 446.
8. Dingxin Zhao, *The Confucian-Legalist State: A New Theory of Chinese History* (New York: Oxford University Press), 8.
9. Johann P. Arnason, "Rehistoricizing the Axial Age," in *The Axial Age and its Consequences*, 354.

10. Robert Bellah, "What Is Axial About the Axial Age?" *European Journal of Sociology* 46, no. 1 (2005): 80.
11. Arnaldo Momigliano, *Alien Wisdom: The Limits of Hellenization* (Cambridge: Cambridge University Press, 1975), 8–9, quoted in Bellah, "What Is Axial About the Axial Age?" 72.
12. Nicolas Baumard et al. "Increased Affluence Explains the Emergence of Ascetic Wisdoms and Moralizing Religion," *Current Biology* 25, no. 1 (January 2015): 10. http://www.sciencedirect.com/science/article/pii/S0960982214013724. Thanks to John Boy for this valuable reference.
13. Eugene Halton, *From the Axial Age to the Moral Revolution: John Stuart-Glennie, Karl Jaspers, and a New Understanding of the Idea* (New York: Palgrave MacMillan, 2014).
14. See Dieter Metzler, "A. H. Anquetil-Duperron (1731–1805) und das Konzept der Achsenzeit," in *Through Travellers' Eyes: European Travellers and the Iranian Monuments*, ed. H. Sancisi-Weerdenburg and J. W. Drijvers (Leiden: Nederlands Instituut voor het Nabije Oosten, 1991); and Ernst von Lasaulx, *Neuer Versuch einer Philosophie der Geschichte* (Munich: n.p., 1856).
15. Halton, *From the Axial Age to the Moral Revolution*, 52.
16. Weber, *Economy and Society*, 1: 441.
17. Alfred Weber, *Kulturgeschichte als Kultursoziologie* (Leiden: Sijthoff, 1935), 7–9.
18. Jaspers, *Origin and Goal of History*, 18.
19. Bellah, "What Is Axial about the Axial Age?" 88.
20. Baumard et al., "Increased Affluence Explains," 10.
21. Eisenstadt, "Axial Age," 201.
22. Baumard et al., "Increased Affluence Explains," 12–13.
23. Lizzie Wade, "Wealth May Have Driven the Rise of Today's Religions," *Science*, December 11, 2014, http://www.sciencemag.org/news/2014/12/wealth-may-have-driven-rise-today-s-religions.
24. Weber, *Economy and Society*, 1: 512.
25. As Weber says, "The materialist conception of history is not a cab to be taken at will; it does not stop short of the promoters of

revolution" ("Politics as a Vocation," in *From Max Weber*, ed. Hans Gerth and C. Wright Mills [New York: Oxford University Press, 1946], 125).

26. Weber, *Economy and Society*, 1: 442, 447.
27. Stanley Wolpert, *A New History of India*, 6th ed. (New York: Oxford University Press, 2000), 43.
28. Ibid., 32.
29. See Somini Sengupta, *The End of Karma: Hope and Fury Among India's Young* (New York: Simon & Schuster, 2016).
30. Wolpert, *New History of India*, 42.
31. Ibid., 38.
32. Ibid., 46.
33. Ibid., 47.
34. Weber, *Economy and Society*, 1: 562.
35. Robert Bellah, *Religion in Human Evolution: From the Paleolithic to the Axial Age* (Cambridge, MA: Belknap/Harvard University Press, 2011), 489.
36. Edward L. Farmer et al., *10,000 B.C to 1850*, vol. 1, *History of Civilization in Asia* (Reading, MA: Addison-Wesley, 1977), 104.
37. William Dalrymple, *Nine Lives: In Search of the Sacred in Modern India* (New York: Alfred A. Knopf, 2010), quoted in David Shulman, "Living in India's Spirit World," *New York Review of Books*, October 28, 2010, http://www.nybooks.com/articles/2010/10/28/living-indias-spirit-world/.
38. Randall Collins, *The Sociology of Philosophies: A Global Theory of Intellectual Change* (Cambridge, MA: Belknap/Harvard University Press, 1998), 205.
39. On this point, see Philip Gorski, *The Disciplinary Revolution: Calvinism and the Rise of the State in Early Modern Europe* (Chicago: University of Chicago Press, 2003), 21.
40. Farmer et al., *Comparative History of Civilizations in Asia*, 1: 96.
41. Bellah writes that the word Israel means "El rules," and notes that this would also buttress the notion of El as the Israelites' chief god before the rise of Yahweh to that position (*Religion in Human Evolution*, 287).

42. Diarmaid MacCulloch, *Christianity: The First Three Thousand Years* (New York: Viking, 2009), 50.
43. Ibid., 58.
44. Ibid., 57.
45. Bellah, *Religion in Human Evolution*, 322.
46. M. I. Finley, *The Ancient Greeks* (New York: Penguin, 1963), 48.
47. Quoted in Ibid., 26.
48. Max Weber, *The Protestant Ethic and the Spirit of Capitalism and Other Writings*, ed. Peter Baehr and Gordon Wells (New York: Penguin, 2002), 73.
49. MacCulloch, *Christianity*, 70.
50. Max Weber, "The Social Psychology of the World Religions," in *From Max Weber: Essays in Sociology*, ed. H. H. Gerth and C. Wright Mills (New York: Oxford University Press, 1945), 293.
51. Max Weber, *The Religion of China: Confucianism and Taoism*, trans. and ed. Hans H. Gerth (New York: Oxford University Press, 1951).
52. Zhao, *Confucian-Legalist State*, 15.
53. Weber, *Religion of China*, 236.
54. Bellah, *Religion in Human Evolution*, 476.
55. Ibid., 279.
56. Jan Assmann, "Cultural Memory and the Myth of the Axial Age," in *The Axial Age and its Consequences*, 401.

2. The Material Axial Age

1. Ian Morris, *Why the West Rules—For Now* (New York: FSG, 2010), 157.
2. Ibid., 353.
3. Alfred Weber, *Kulturgeschichte als Kultursoziologie*, 8.
4. Jacques Le Goff, *Must We Divide History Into Periods?* trans. M. B. DeBevoise (New York: Columbia University Press, 2015), 100.

5. Fernand Braudel, *The Structures of Everyday Life*, vol. 1, *Civilization and Capitalism 15th-18th Century* (Berkeley: University of California Press, 1992), 70.
6. Edmund Burke III, "Toward a Comparative History of the Modern Mediterranean, 1750–1919," *Journal of World History* 23, no. 4: 912–13.
7. Carlo Cipolla, *Before the Industrial Revolution: European Society and Economy, 1000–1700*, 3d ed. (New York: W. W. Norton, 1993), 100.
8. Robert J. Gordon, *The Rise and Fall of American Economic Growth* (Princeton, NJ: Princeton University Press, 2016), 2.
9. This is the fundamental premise of Angus Deaton, *The Great Escape: Health, Wealth, and the Origins of Inequality* (Princeton, NJ: Princeton University Press, 2013).
10. Morris, *Why the West Rules-For Now*, 161.
11. Quoted in Keith Tribe, "Introduction," in Koselleck, *Futures Past: On the Semantics of Historical Time*, trans. Keith Tribe (New York: Columbia University Press, 2004), xiv and xvi.
12. Jürgen Osterhammel, *The Transformation of the World: A Global History of the Nineteenth Century* (Princeton, NJ: Princeton University Press, 2014 [2009]), 58–62.
13. P. J. Crutzen and E. F. Stoermer, "The Anthropocene," *IGBP Global Change Newsletter*, May 2000, 17–18. See also Will Steffen et al., "The Anthropocene: Conceptual and Historical Perspectives," *Philosophical Transactions of the Royal Society A*, January 2011, DOI:10.1098/rsta.2010.0327. And see also Colin N. Waters et al., "The Anthropocene Is Functionally and Stratigraphically Distinct from the Holocene," *Science*, January 8, 2016, http://science.sciencemag.org/content/351/6269/aad2622.full.
14. Will Steffen et al., "The Trajectory of the Anthropocene: The Great Acceleration," *The Anthropocene Review* 2, no. 1 (2015): 82, DOI:10.1177/2053019614564785.
15. For a discussion of the controversies over the dating of the "Anthropocene" and its significance, see Katrina Forrester, "The

Anthropocene Truism," *The Nation*, May 12, 2016, http://www.thenation.com/article/the-anthropocene-truism/.

16. See Robert William Fogel, *The Escape from Hunger and Premature Death, 1700–2100: Europe, America, and the Third World* (New York: Cambridge University Press, 2004); Steven Pinker, *The Better Angels of Our Nature* (New York: Viking, 2011); and Morris, *War! What Is it Good For?*

17. See Arno Mayer, *The Persistence of the Old Regime: Europe to the Great War* (New York: Pantheon, 1981); and Osterhammel, *Transformation of the World*, 643.

18. Morris, *Why the West Rules—For Now*, 144.

19. Deaton, *Great Escape*, 9.

20. Pomeranz, *Great Divergence*.

21. Morris, *Why the West Rules-For Now*, 502. See also Jared Diamond, *Guns, Germs, and Steel: The Fates of Human Societies* (New York: W. W. Norton, 1999).

22. Daron Acemoglu and James Robinson, *Why Nations Fail: The Origins of Power, Prosperity, and Poverty* (New York: Random House, 2012).

23. Osterhammel, *Transformation of the World*, 643.

24. Simon Kuznets, *Modern Economic Growth: Rate, Structure, and Spread* (New Haven, CT: Yale University Press, 1967 [1966]).

25. James Lee and Wang Feng, *One Quarter of Humanity* (Cambridge, MA: Harvard University Press, 1999), 3.

26. "U.S and World Population Clock," infographic, United States Census Bureau, http://www.census.gov/popclock/.

27. See, for example, Gwynn Guilford, "Thailand's Joining Japan and Korea as One of the 'Old Men' of Asia," *Quartz*, August 23, 2013, http://qz.com/118201/thailands-joining-japan-and-korea-as-one-of-the-old-men-of-asia/

28. See Amartya Sen, "Women's Progress Outdid China's One-Child Policy," *New York Times*, November 2, 2015, http://www.nytimes.com/2015/11/02/opinion/amartya-sen-womens-progress-outdid-chinas-one-child-policy.html.

29. See Kate Kelland, "Global Life Expectancy Rises, But People Live Sicker Longer," *Reuters*, August 27, 2015, http://www.reuters.com/article/2015/08/27/us-health-longevity-idUSKCN0QV2JL20150827.
30. Mark Nathan Cohen, *Health and the Rise of Civilization*.
31. Robert William Fogel, *The Escape from Hunger and Premature Death, 1700–2100: Europe, America, and the Third World* (New York: Cambridge University Press, 2004), 40.
32. Deaton, *Great Escape*, 247.
33. David Cutler and Grant Miller, "The Role of Public Health Improvements in Health Advances: The Twentieth-Century United States," *Demography* 42, no. 1 (December 2005): 3.
34. Tim Brewer and Yolana Pringle, "Beyond Bazalgette: 150 Years of Sanitation," *The Lancet* 386, no. 9989 (July 2015): 128–29, DOI:10.1016/S0140–6736(15)61231–4.
35. Deaton, *Great Escape*, 93.
36. Gretchen Newby et al., "The Path to Eradication: A Progress Report on the Malaria-Eliminating Countries," *The Lancet* 387, no. 10029 (April 23, 2016): 1775–84. http://www.thelancet.com/journals/lancet/article/PIIS0140–6736(16)00230–0/fulltext?elsca1=etoc&elsca2=email&elsca3=0140–6736_20160423_387_10029_&elsca4=Public%20Health%7CInfectious%20Diseases%7CHealth%20Policy%7CInternal%2FFamily%20Medicine%7CGeneral%20Surgery%7CLancet
37. Donald G. McNeil, Jr., "A Milestone in Africa: No Polio Cases in a Year," *New York Times*, August 11, 2015, http://www.nytimes.com/2015/08/12/health/a-milestone-in-africa-one-year-without-a-case-of-polio.html?_r=0.
38. "Pakistan Arrests Parents for Opposing Polio Vaccine," *BBC News*, March 2, 2015, http://www.bbc.com/news/world-asia-31703835.

39. Sabrina Tavernise, "Life Expectancy Increases Around the World," *New York Times*, December 13, 2012, http://www.nytimes.com/2012/12/14/health/worlds-population-living-longer-new-report-suggests.html?_r=0.
40. Claudia Goldin and Robert Margo, "The Great Compression: The Wage Structure in the United States at Mid-Century," *Quarterly Journal of Economics* 107 (February 1992): 1–34.
41. Branko Milanovic, "Global Income Inequality By the Numbers: In History and Now—An Overview," *World Bank Development Research Team, Policy and Inequality Team*, November 2012, 7–8; http://www-wds.worldbank.org/external/default/WDSContentServer/WDSP/IB/2012/11/06/000158349_20121106085546/Rendered/PDF/wps6259.pdf.
42. See Steven Radelet, *The Great Surge: The Ascent of the Developing World* (New York: Simon & Schuster, 2015), 65–71.
43. Sarah Percy, *Mercenaries: The History of a Norm in International Relations* (New York: Oxford University Press, 2007).
44. See Pinker, *Better Angels*, 60–63.
45. See Stephen Mennell, *The American Civilizing Process* (Malden, MA: Blackwell, 2006).
46. The basic argument is that of Norbert Elias, *The Civilizing Process* (Malden, MA: Blackwell, 2000 [1939]).
47. James J. Sheehan, *Where Have All the Soldiers Gone? The Transformation of Modern Europe* (New York and Boston: Houghton Mifflin, 2008).
48. Elizabeth Kolbert, *The Sixth Extinction: An Unnatural History* (New York: Henry Holt, 2014).

3. The Mental Axial Age

1. Daniel Bell, *The Coming of Post-Industrial Society: A Venture in Social Forecasting* (New York: Basic Books, 1973).

2. David Card and John E. Dinardo, "Skill-Biased Technical Change and Rising Wage Inequality: Some Puzzles and Problems," *Journal of Labor and Economics* 20, no. 4 (2002).
3. See Robert J. Gordon, *The Rise and Fall of American Economic Growth* (Princeton, NJ: Princeton University Press, 2016), 613ff.
4. Branko Milanovic, *Global Inequality: A New Approach for the Age of Globalization* (Cambridge, MA: Belknap/Harvard University Press, 2016), 23.
5. P. W. Singer, *Wired for War: The Robotics Revolution and Conflict in the 21st Century* (New York: Penguin, 2009), 101.
6. Farhad Manjoo, "For the New Year, Let's Resolve to Improve Our Tech Literacy," *New York Times*, December 23, 2015, http://www.nytimes.com/2015/12/24/technology/for-the-new-year-lets-resolve-to-improve-our-tech-literacy.html?_r=1.
7. Eric Brynjolfsson and Andrew McAfee, *The Second Machine Age: Work, Progress, and Prosperity in a Time of Brilliant Technologies* (New York: W. W. Norton, 2014), 41.
8. Quoted in Gordon, *Rise and Fall of American Growth*, 17. The other points regarding the pessimistic view of the new technologies are the central theses of his book.
9. "The Trouble with GDP," *The Economist,* April 30, 2016, 24.
10. See Brynjolfsson and McAfee, *Second Machine Age*, 11.
11. David H. Autor, "Why Are There Still so Many Jobs? The History and Future of Workplace Automation," *Journal of Economic Perspectives* 29, no. 3 (summer 2015), https://www.aeaweb.org/articles?id=10.1257/jep.29.3.3. See also John Markoff, "The End of Lawyers? Not So Fast," *New York Times*, January 4, 2016, http://bits.blogs.nytimes.com/2016/01/04/the-end-of-work-not-so-fast/.
12. John Markoff, "How Tech Giants are Devising Real Ethics for Artifical Intelligence," *New York Times*, September 2, 2016, http://www.nytimes.com/2016/09/02/technology/artificial-intelligence-ethics.html?_r=0.
13. Robert C. Allen, *Global Economic History: A Very Short Introduction* (New York: Oxford University Press, 2011), 39.

14. Michael Specter, "The Gene Hackers," *The New Yorker,* November 16, 2015, 52.
15. Singer, *Wired for War*, 72.
16. Reuters, "Device Harnessing Thoughts Allows Quadriplegic to Use his Hands," *New York Times* April 13, 2016, http://www.nytimes.com/reuters/2016/04/13/us/13reuters-science-paralysis.html.
17. John Markoff, "Pentagon Turns to Silicon Valley for Edge in Artificial Intelligence," *New York Times* May 11, 2016, http://www.nytimes.com/2016/05/12/technology/artificial-intelligence-as-the-pentagons-latest-weapon.html?_r=0.
18. Johan Rockström, Gunhild Anker Stordalen, Richard Horton, "Acting in the Anthropocene," *The Lancet* 387, no. 10036 (June 11, 2016), http://ac.els-cdn.com/S014067361630681X/1-s2.0-S014067361630681X-main.pdf?_tid=24ddcb5a-32f9-11e6-a463-00000aacb360&acdnat=1465995812_53434ad68623c062264bc2c7b36223fa
19. Justin Gillis, "Business Leaders Back 'Net Zero' Target," *New York Times,* December 7, 2015, http://www.nytimes.com/interactive/projects/cp/climate/2015-paris-climate-talks/business-leaders-set-ambitious-new-goals.
20. Michael Slezak, "World's Carbon Dioxide Concentration Teetering on the Point of No Return," *The Guardian,* May 10, 2016, http://www.theguardian.com/environment/2016/may/11/worlds-carbon-dioxide-concentration-teetering-on-the-point-of-no-return?utm_source=esp&utm_medium=Email&utm_campaign=GU+Today+USA+-+morning+briefing+2016&utm_term=171647&subid=9554953&CMP=em a_a-morning-briefing_b-morning-briefing_c-US_d-1.
21. Joel Mokyr, "Is Technological Progress a Thing of the Past?" *VOX*, September 8, 2013, http://voxeu.org/article/technological-progress-thing-past.
22. Ralf Fücks, *Green Growth, Smart Growth: A New Approach to Economics, Innovation, and the Environment* (New York: Anthem, 2015).

23. See John Torpey and Saskia Hooiveld, "Warfare Without Warriors? Changes in Contemporary Warfare and the Decline of the Citizen-Soldier," in *Transformations of Warfare in the Modern World*, ed. John Torpey and David Jacobson (Philadelphia, PA: Temple University Press, 2016), 1–19.
24. Therése Pettersson and Peter Wallensteen, "Armed Conflict 1946–2014," *Journal of Peace Research* 52, no. 4 (2015): 537.
25. Andreas Wimmer, *Waves of War: Nationalism, State Formation, and Ethnic Exclusion in the Modern World* (New York: Cambridge University Press, 2013), 27.
26. Mary Kaldor, *New and Old Wars: Organized Violence in a Global Era*, 3d ed. (Stanford, CA: Stanford University Press, 2012).
27. Singer, *Wired for War*, 58.
28. Cecilia Kang, "More Than 180,000 Drone Users Registered in F.A.A. Database," *New York Times* January 6, 2016, http://bits.blogs.nytimes.com/2016/01/06/more-than-180000-drone-users-registered-in-f-a-a-database/.
29. Mark Mazzetti and David E. Sanger, "Security Chief Says U.S. Would Retaliate Against Cyberattacks," *New York Times*, March 12, 2013, http://www.nytimes.com/2013/03/13/us/intelligence-official-warns-congress-that-cyberattacks-pose-threat-to-us.html?pagewanted=all.
30. David E. Sanger, "In Cyberspace, New Cold War," *New York Times*, February 24, 2013, http://www.nytimes.com/2013/02/25/world/asia/us-confronts-cyber-cold-war-with-china.html?pagewanted=all&_r=0; see also Jane Perlez, "U.S. and China Put Focus on Cybersecurity," *New York Times*, April 22, 2013, http://www.nytimes.com/2013/04/23/world/asia/united-states-and-china-hold-military-talks-with-cybersecurity-a-focus.html.
31. Mike Hoffman, "Cyber Is Likely Winner of 2015 Budget," *Defensetech*, February 24, 2014, http://defensetech.org/2014/02/24/cyber-is-likely-winner-of-2015-budget/.
32. See, for example, Ariel Colonomos, "Precision Warfare and the Case for Symmetry: Targeted Killings and Hostage-Taking,"

in *Transformations of Warfare in the Modern World*, ed. John Torpey and David Jacobson (Philadelphia, PA: Temple University Press, 2016). See also William J. Broad and David E. Sanger, "As U.S. Modernizes Nuclear Weapons, 'Smaller' Leaves Some Uneasy," *New York Times,* January 11, 2016, http://www.nytimes.com/2016/01/12/science/as-us-modernizes-nuclear-weapons-smaller-leaves-some-uneasy.html?_r=0; and Markoff, "Pentagon Turns to Silicon Valley."

33. For a dramatization of the way drone strike decisions are made among military, legal, and political actors today, see the excellent film *Eye in the Sky* (2015).
34. J. Oeppen and J. Vaupel, "Broken Limits to Life Expectancy," *Science*, May 10, 2002, http://www.econ.ku.dk/okocg/VV/VV-Economic%20Growth/articles/artikler-2006/Broken-limits-to-life-expectancy.pdf.
35. Robert William Fogel, *The Escape from Hunger and Premature Death*, 108–11.
36. Simon Romero, "An Exploding Pension Crisis Fuels Brazil's Political Turmoil," *New York Times,* October 20, 2015, http://www.nytimes.com/2015/10/21/world/americas/brazil-pension-crisis-mounts-as-more-retire-earlier-then-pass-benefits-on.html.
37. Natasha Singer and Mike Isaac, "Mark Zuckerberg's Philanthropy Uses L.L.C. For More Control," *New York Times,* December 2, 2015, http://www.nytimes.com/2015/12/03/technology/zuckerbergs-philanthropy-uses-llc-for-more-control.html.
38. Bryan S. Turner, *Can We Live Forever? A Sociological and Moral Inquiry* (New York: Anthem, 2009).
39. Global Burden of Disease Study 2013 Collaborators [Theo Vos et al.], "Global, Regional, and National Incidence, Prevalence, and Years Lived with Disability for 301 Acute and Chronic Diseases and Injuries in 188 Countries, 1990–2013: A Systematic Analysis for the Global Burden of Disease Study 2013," *The Lancet*, June 8, 2015, 1–58, http://www.thelancet.com/pdfs/journals/lancet/PIIS0140-6736(15)60692-4.pdf.

40. John R. Beard et al., "World Report on Ageing and Health: A Policy Framework for Healthy Ageing," *The Lancet* 387, no. 2150 (2016).
41. Sabrina Tavernise, "Global Diabetes Rates are Rising as Obesity Spreads," *New York Times*, June 8, 2015, http://www.nytimes.com/2015/06/08/health/research/global-diabetes-rates-are-rising-as-obesity-spreads.html?_r=0.
42. John R. Beard et al., "World Report on Ageing and Health: A Policy Framework for Healthy Ageing," *The Lancet* 387, no. 2149 (2016).
43. Joseph Dieleman et al., "National Spending on Health By Source Between 2013 and 2040," *The Lancet* 387, (2016).
44. See Neil G. Bennett and Bo Chen, "Forecasting Civil Conflict," paper presented at the 2013 meeting of the Population Association of America; see also the papers collected in Jack Goldstone et al., eds., *Political Demography: How Population Changes Are Reshaping International Security and National Politics* (New York: Oxford University Press, 2013).
45. World Health Organization, *World Report on Ageing and Health* (Geneva: WHO, 2015).
46. T. H. Marshall, "Citizenship and Social Class," in *Class, Citizenship, and Social Development* (Garden City, NY: Anchor/Doubleday, 1963), 78.
47. *2013 Report Card for America's Infrastructure* (American Society of Civil Engineers, 2013), http://www.infrastructurereportcard.org/a/documents/2013-Report-Card.pdf. For a discussion, see Elizabeth Drew, "A Country Breaking Down," *New York Review of Books*, February 25, 2016, http://www.nybooks.com/articles/2016/02/25/infrastructure-country-breaking-down/.
48. David Sedlak, *Water 4.0: The Past, Present, and Future of the World's Most Vital Resource* (New Haven, CT: Yale University Press, 2014), 87.
49. See World Health Organization, "Air Pollution Levels Rising in Many of the World's Poorest Cities," news release, May 12, 2016, http://who.int/mediacentre/news/releases/2016/air-pollution-rising/en/.

50. John Komlos and Marieluise Baur, "From the Tallest to (One of) the Fattest: The Enigmatic Fate of the American Population in the 20th Century," *Economics and Human Biology* 2 (2004): 57. http://user37685.vs.easily.co.uk/wp/wp-content/uploads/2013/10/komlos-baur-2004.pdf.

51. Steven H. Woolf and Laudan Aron, eds., *U.S. Health in International Perspective: Shorter Lives, Poorer Health* (Washington, DC: National Academy of Sciences, 2013), 1.

52. S. Jay Olshansky et al., "Differences in Life Expectancy Due to Race and Educational Differences are Widening, and Many May Not Catch Up," *Health Affairs* 31, no. 8 (2012): 1806, http://content.healthaffairs.org/content/31/8/1803.full.pdf+html.

53. Raj Chetty et al., "The Association Between Income and Life Expectancy in the United States, 2000–2014," *Journal of the American Medical Association*, April 16, 2016, http://jama.jamanetwork.com/article.aspx?articleid=2513561.

54. Anne Case and Angus Deaton, "Rising Mortality and Morbidity in Midlife Among White Non-Hispanic Americans in the 21st Century," *Proceedings of the National Academy of Sciences* 112, no. 49 (December 8, 2015): 15081, http://www.pnas.org/content/112/49/15078.full.pdf.

55. Sally Curtin, Margaret Warner, and Holly Hedegaard, "Increase in Suicide in the United States, 1999–2014," *National Center for Health Statistics*, no. 241 (April 2016), http://www.cdc.gov/nchs/products/databriefs/db241.htm

56. S. Jay Olshansky et al., "Differences in Life Expectancy Due to Race and Educational Differences are Widening, and Many May Not Catch Up," *Health Affairs* 31, no. 8 (2012): 1808, http://content.healthaffairs.org/content/31/8/1803.full.pdf+html.

57. J. Currie and H. Schwandt, "Inequality in Mortality Decreased Among the Young While Increasing For Older Adults, 1990–2010," *Science,* April 2016, http://science.sciencemag.org/content/early/2016/04/20/science.aaf1437.

58. Fücks, *Green Growth, Smart Growth*, 99.

59. Amartya Sen, *Development as Freedom* (New York: Anchor, 1999), Ch. 7.
60. Brynjolfsson and McAfee, *Second Machine Age*, 256.
61. See Sherry Turkle, *Alone Together: Why We Expect More from Technology and Less from Each Other* (New York: Basic, 2011); and Sherry Turkle, *Reclaiming Conversation: The Power of Talk in a Digital Age* (New York: Penguin, 2015).
62. See their website, http://www.helpage.org/
63. See John Markoff, "As Aging Populations Grow, So Do Robotic Health Aids," *New York Times*, December 8, 2015, http://www.nytimes.com/2015/12/08/science/as-aging-population-grows-so-do-robotic-health-aides.html?_r=0.
64. Fogel, *Escape from Hunger*, 74–80.
65. For a sampling of the range of views on the idea, see Joel Rogers, Joshua Cohen, and Philippe van Parijs, eds., *What's Wrong with a Free Lunch?* (Boston: Beacon Press, 2001). After a number of articles on the topic by *New York Times* economics columnist Eduardo Porter and tech columnist Farhad Manjoo, pieces appeared roughly simultaneously in *The New Yorker* (http://www.newyorker.com/magazine/2016/06/20/why-dont-we-have-universal-basic-income; June 20, 2016—for) and *The Economist* (http://www.economist.com/news/leaders/21699907-proponents-basic-income-underestimate-how-disruptive-it-would-be-basically-flawed; June 4, 2016—against).
66. See James Surowiecki, "Why Don't We Have a Basic Income?" *The New Yorker*, June 20, 2016, http://www.newyorker.com/magazine/2016/06/20/why-dont-we-have-universal-basic-income. And see also Jathan Sadowski, "Why Silicon Valley Is Embracing Universal Basic Income," *The Guardian*, June 22, 2016, https://www.theguardian.com/technology/2016/jun/22/silicon-valley-universal-basic-income-y-combinator.
67. Raffi Khatchadourian, "The Doomsday Invention," *The New Yorker*, November 23, 2015, 74.

68. See Robert M. Geraci, *Apocalyptic AI: Visions of Heaven in Robotics, Artificial Intelligence, and Virtual Reality* (New York: Oxford University Press, 2010).
69. Adam Davidson, "Saving the World, Startup Style," *New York Times Magazine*, November 22, 2015.
70. Coral Davenport, "Can Economies Rise as Emissions Fall? The Evidence Says Yes," *New York Times*, April 5, 2016, http://www.nytimes.com/2016/04/06/upshot/promising-signs-that-economies-can-rise-as-carbon-emissions-decline.html.
71. Stephane Hallegatte et al., *Shock Waves: Managing the Impacts of Climate Change on Poverty*. Climate Change and Development Series. Washington, DC: World Bank. Doi:10.1596/978-1-4648-0673-5. License: Creative Commons Attribution CC BY 3.0 IGO.
72. John Markoff, *Machines of Loving Grace: The Quest for Common Ground Between Humans and Robots* (New York: Ecco/HarperCollins, 2015), 54.
73. Pope Francis, *Laudato Si': On Care for Our Common Home* (2015), available at http://w2.vatican.va/content/dam/francesco/pdf/encyclicals/documents/papa-francesco_20150524_enciclica-laudato-si_en.pdf.
74. Bill McKibben, "The Pope and the Planet," *New York Review of Books*, August 13, 2015, http://www.nybooks.com/articles/archives/2015/aug/13/pope-and-planet/.
75. David Gelles, "How Producing Clean Power Turned Out to Be a Messy Business," *New York Times*, August 13, 2016, http://www.nytimes.com/2016/08/14/business/energy-environment/how-producing-clean-power-turned-out-to-be-a-messy-business.html?_r=0.
76. Associate Press, "Pope Calls for a New Work of Mercy: Care for the Environment," *Los Angeles Times*, September 1, 2016, http://www.latimes.com/world/.

Index

African Americans: education levels, 69–70; inequality and, 76
African religions, 31
Agni (god), 14
agriculture and food production, 7–8, 34–35, 57, 70–71
ahimsa, 18, 21
air quality and pollution, 67
Allen, Robert, 55
American indigenous religions, 31
American Society of Civil Engineers (ASCE), 66–67
Anquetil-Duperron, Abraham Hyacinthe, 10
"Anthropocene" era, 37–38, 51, 57–58
Aristotle, 27–28
Arnason, Johann, 9
artificial intelligence, 2, 3, 4, 50, 56–57, 73–74
Assmann, Jan, 31–32
Assyrians, 23–24
Autor, David, 55
Axial Age thesis, vii–viii, 1, 8–11, 31

Baal (god), 24
Baumard, Nicholas, 9, 11–12, 52
Bell, Daniel, 51
Bellah, Robert, vii–viii, ix, 4, 9–10, 11, 17–18, 23–24, 30–31
Benjamin, Walter, 43

Bezos, Jeff, 60
Blake, William, 43
Bostrom, Nick, 74
Bourdieu, Pierre, ix
brahman caste, 14–16
Braudel, Fernand, 34
Brazil: in BRICS, 46; life expectancy in, 62
Brynjolfsson, Eric, 54, 71
Buddha and Buddhism, viii, 1, 2, 8, 13, 19–22
Burckhardt, Jacob, 2
Burke, Terry, 34

cancer, 45, 63
Case, Anne, 69
China, vii, 8–9; Confucianism in, 28–31; cyber weapons, 61; economic growth, 42, 46; "mandate of heaven" idea, 30; population growth, 41–42
cholera, 44
Christianity, 10, 13, 18, 25, 27–28, 33, 34, 77–78; Protestant Reformation, 20–21, 27–28, 29–30
cities, rise of, 7–8
climate change and global warming, 4, 5, 37–38, 51, 58, 76–78
communications technology, 2, 3, 58, 71–72, 76

computers and computerization, 3, 51, 54–56, 60–62, 74
Condorcet, Marquis de, x
Confucius and Confucianism, viii, 1, 2, 13, 28–31
Crutzen, Paul, 37
cyber warfare, 61–62
cyborgs, 56–57

Deaton, Angus, 39, 44, 69
demographic transition, 41–42
diabetes, 45–46
Diamond, Jared, 39
digital natives, 50–51, 72
drones, 60

economic issues, 3, 33, 35–43, 51–56, 73, 77
Ehrlich, Paul, 71
Eisenstadt, Shmuel N., 7, 8, 10, 12
El (god), 23–24
elderly populations, 63–66, 72–73
Elijah, 24–25
employment issues, 4, 5–6, 51–52, 54–55, 72–73
environmental issues, 3, 4–5, 48–49, 51, 77–78. *See also* climate change and global warming
"ethical" prophets, 13–14, 17, 24–25
"exemplary" prophets, 13–14, 17, 28–29

fertility rates, 41–42
Fogel, Robert, 42, 62, 73
fossil fuels, 3, 37–38, 51
Francis, Pope, 77–78
French Revolution, 47
Freud, Sigmund, 43
Friedman, Thomas, 46
Fücks, Ralf, 58

Gandhi, Mahatma, 18
GDP as measurement, 55

genetic engineering, 50, 56
global warming. *See* climate change and global warming
globalization, 51–52, 74–76
Gordon, Robert J., 35, 54
"gray peace dividend," 64–65
Great Acceleration, 37–38, 52
Great Compression, 46, 51, 52, 54
Greek culture, government, and art, 25–26
Greek philosophy, vii, viii, 1, 3, 8, 27–28
Greek religion, 26–27
Gulf War, 60

Habermas, Jürgen, ix
Halton, Eugene, 10–11, 32
health care and nutrition, 3, 38, 41, 42, 43–46, 57, 62–70, 71
heart disease, 63
Hegel, Georg Wilhelm Friedrich, 1
HelpAge International, 73
Herodotus, 27
Hesiod, 27
Hinduism, viii, 13, 14–18, 21–22
Homer, 27
homicide rates, 47
hunter-gatherers, 7–8

Ibn Khaldûn, 7
income inequality, 3, 46, 48, 51–52, 54–55, 68
India, vii, 8; caste system in, 15–16, 18; conflict with Pakistan, 59; economic growth, 42, 46; population growth, 41–42; religion in, 14–22
Industrial Revolution, 34–35, 37–39, 48–49, 56
information technology, 2, 3, 50, 58, 76
infrastructure maintenance, 66–67
Inglehart, Ronald, 11

Isaiah, 25
Islam, 10, 13, 18, 33–34
Israel, ancient, 22–23

Jainism, 18–19, 20, 21
Jaspers, Karl, 1, 8–10, 11, 31
Judaism, viii, 1, 2, 8, 13, 17, 22–25

karma, 17, 18–19, 20, 22
Koselleck, Reinhard, 3, 35–37
kshatriya caste, 15

Lasaulx, Ernst von, 10
Le Goff, Jacques, 34
life expectancy, 3, 38, 41, 42, 44, 45–46, 48, 62–66, 68–69

Maccabean revolt, 28
MacCulloch, Diarmaid, 23–24
Magadha dynasty, 15–16
Mahabharata epic, 16
Mahavira, 18, 19
malaria, 45
Malthus, Thomas, 70–71
Manjoo, Farhad, 53
Mann, Michael, viii
Markoff, John, 77–78
Marshall, T. H., 66
Marx, Karl, 4, 5, 13
Material Axial Age (second), viii, ix–x, 33–49; characteristics of, 1–3, 35–36; demographic changes, 41–42; energy regime, 3, 37–38, 56; global economic inequality in, 3, 46; health advances, 42, 43–46; mode of thought, 4; *Sattelzeit* period, 3, 35–37; social development, 36, 38–40; violent death rates, 47–48, 71
McAfee, Andrew, 54, 71
McKibben, Bill, 78
Mental Axial Age (third), viii, ix–x, 50–78; challenges of, x, 76–78; characteristics of, 1–3; economic issues, 51–56, 73, 77; energy regime, 3–4, 51, 57–58; health care and life expectancy, 62–70; melding of ideas from earlier ages, 4–5; mode of thought, 4–5, 50; nostalgia and, 52; population growth, 71; technological breakthroughs, 2, 3, 4, 50, 51–57, 71–72, 74–76; violent death risk, 71; warfare, 58–62
Mexico, obesity in, 67
Milanovic, Branko, 46
mindfulness meditation, 72
moksha, 16–17, 20
Mokyr, Joel, 58
Momigliano, Arnaldo, 9
"Moore's Law," 53
Moral Axial Age (first), viii, ix–x, 7–32; causes of, 11–13; characteristics of, 1–3, 10–11; energy regime, 3; idea of transcendence in, 2–3, 8–9; mode of thought, 4, 9–10, 11–12; religion and philosophy in, 12–32
Morris, Ian, 33, 38–39
mortality rates, 41–42, 44–45, 63, 69–70
Muhammad, 13

nirvana, 20
nuclear weapons, 58–59
nutrition. *See* health care and nutrition

obesity, 45–46, 67
Oepen, J., 62
Olshansky, S. Jay, 68, 69–70
Osterhammel, Jürgen, 35–36, 40

Pakistan: conflict with India, 59; polio in, 45
Palestine, ancient, 8, 17, 22–23

panzoonism, 10
Parsons, Talcott, ix, 10
Piketty, Thomas, 46
Plato, 27–28
Polanyi, Karl, 37–38
polio, 45
polis, 25–26
Pomeranz, Kenneth, 39
population growth, world, 3, 41–42, 48, 70–71
post-materialism, 11
poverty reduction, 42–43, 76
privatization issues, 75–76
progress, idea of, x, 1, 43
prophecy, forms of, 13–14
"Protestant ethic," 40

religion, vii, viii, 2–3, 8, 12–32, 72. *See also specific religions*
renewable energy sources, 3–4, 57–58
Rig Veda, 14–18
robotics, 2, 3, 50, 60, 75
Roman Empire, 33
Russia: in BRICS, 46; life expectancy in, 62

Sahlins, Marshall, 7
samsara, 17–18
sanitation infrastructure, 43–45, 66–67
Sattelzeit, 3, 35–37
self-understanding and self-reflection, 1, 4, 9–10, 15, 30, 35
Sen, Amartya, 55, 71
shudra caste, 15–16
Simon, Julian, 71
Singer, P. W., 53, 60
slavery and slave labor, 36, 40
"smart" technologies, 4, 50, 51
Smith, Adam, 4
Socrates, 8

Solomon, King, 22–23
Solow, Robert, 54
South Africa: in BRICS, 46
Soviet Union, collapse of, 46
Stoermer, Eugene, 37
Stuart-Glennie, John, 10–11
Syria, 59–60

Taoism, viii, 29
techno-libertarians, 74–75
techno-optimists, 52–53, 55, 58
techno-pessimists, 53–55
terrorism, 48, 59–60
transcendence, idea of, 2–3, 8–9, 27, 74
tuberculosis, 45
Turkle, Sherry, 71–72
typhoid fever, 44

United Kingdom, sanitation improvements in, 43
United States: cyber weapons of, 61; education levels, 69–70; health care in, 67–68; homicide rates, 48; income inequality in, 46, 51–52, 68; infrastructure challenges, 66; life expectancy, 68–69; obesity in, 67; social safety nets, 67–68; suicide rate, 69; water purification advances, 43, 66–67
universal basic income, 73
"untouchables," 15
Upanishads, 16–17

vaccination campaigns, 43–44, 45
vaishya caste, 15
Vaupel, J., 62
Vedic Brahmanism, 14–18, 20–21, 27
violent death, risk of, 6, 8, 38, 47–48, 71
virtual reality, 50

wars and warfare, 47, 56–57, 58–62
water and milk supply improvements, 43–44, 66–67
Watt, James, 37, 39
Weber, Alfred, 11, 34
Weber, Max, 2, 8, 11, 12–14, 17, 22, 27–28, 29–30, 40
Wimmer, Andreas, 59
women: life expectancy, 68–69; rights and equality for, 36, 38, 76
World Bank, 76

World Health Organization (WHO), 63, 65–66, 67

Xi Jinping, 28

Yahweh (god), 23–24
Y-Combinator, 73

Zero Campaign, 76
Zoroaster and Zoroastrianism, 13, 14
Zuckerberg, Mark, 63

About the Author

JOHN TORPEY is Presidential Professor of Sociology and History and Director of the Ralph Bunche Institute for International Studies at the Graduate Center, City University of New York.